TECHNICAL REPORT

T0302547

Options for Filling Vacant Officer Positions

Roland J. Yardley, Harry J. Thie, M. Wade Markel, Thomas Manacapilli, Joseph Jenkins

Prepared for the Office of the Secretary of Defense

RAND NATIONAL DEFENSE RESEARCH INSTITUTE

The research described in this report was prepared for the Office of the Secretary of Defense (OSD). The research was conducted in the RAND National Defense Research Institute, a federally funded research and development center sponsored by the OSD, the Joint Staff, the Unified Combatant Commands, the Department of the Navy, the Marine Corps, the defense agencies, and the defense Intelligence Community under Contract W74V8H-06-C-0002.

Library of Congress Cataloging-in-Publication Data

Options for filling vacant officer positions / Roland J. Yardley ... [et al.].
p. cm.
Includes bibliographical references and index.
ISBN 978-0-8330-5058-8 (pbk. : alk. paper)
1. United States—Armed Forces—Officers—Supply and demand. I. Yardley, Roland J.

UB413.O65 2010
355.6'1—dc22

2010050309

Published 2010 by the RAND Corporation
1776 Main Street, P.O. Box 2138, Santa Monica, CA 90407-2138
1200 South Hayes Street, Arlington, VA 22202-5050
4570 Fifth Avenue, Suite 600, Pittsburgh, PA 15213-2665
RAND URL: http://www.rand.org/
To order RAND documents or to obtain additional information, contact
Distribution Services: Telephone: (310) 451-7002;
Fax: (310) 451-6915; Email: order@rand.org

Preface

The military services and combatant commanders have officer vacancies on their staffs. That is, funded officer manpower authorizations (or billets) are unfilled. When this occurs, organizational performance and mission readiness can suffer. The RAND Corporation was asked to determine whether the "personnel fill" of funded but vacant military manpower billets could be improved (or at least not degraded) through the use of alternative mixes of military manpower categories, thereby improving organizational performance while possibly also minimizing the cost of manpower. This report reviews particular selected occupations in each of the services, considers extant Department of Defense (DoD) military manpower policy, and assesses alternative mixes of military manpower against criteria developed to determine feasibility. The report should be of interest to the military manpower community. Comments are welcome and may be sent to Roland J. Yardley at Roland_Yardley@rand.org or to Harry J. Thie at Harry_Thie@rand.org.

This research was sponsored by the Office of the Secretary of Defense and conducted within the Forces and Resources Policy Center of the RAND National Defense Research Institute, a federally funded research and development center sponsored by the Office of the Secretary of Defense, the Joint Staff, the Unified Combatant Commands, the Navy, the Marine Corps, the defense agencies, and the defense Intelligence Community.

For more information on the RAND Forces and Resources Center, see http://www.rand.org/nsrd/about/frp.html or contact the Director (contact information is provided on the web page).

Contents

Figures

Tables

Summary

Background and Purpose

The military services and combatant commands (COCOMs) have officer vacancies on their staffs. That is, funded officer manpower authorizations (or billets) are unfilled. Officer vacancies result when an officer with the requisite qualifications is not available to serve in a billet that requires such an officer. When this occurs, organizational performance and mission readiness can suffer.

The services and the COCOMs have reported vacancies in funded officer billets. This has prompted DoD to become more interested in how best to use military manpower to meet mission demands. However, efforts to rectify the issue are complicated by the services' different approaches to meeting these demands. The services use different categories of military manpower to meet DoD mission needs, including the categories of commissioned officer, warrant officer (WO), limited-duty officer (LDO), and enlisted. Service approaches even to what appear to be the same duties can vary, as, for example, when the Air Force uses commissioned officers to pilot unmanned aviation systems while the Army uses enlisted personnel.

In recent decades, three broad trends have reshaped the U.S. military and the work it performs. First, the number of active-duty personnel has diminished to 1.4 million, nearly its lowest number in the post–World War II era. Second, the qualifications of military personnel continue to improve, with the current force being the most qualified in U.S. military history. Third, the nature of military work continues to evolve and has become increasingly technical.

To better understand how to match its changing workforce with its evolving needs, DoD asked RAND to assess possible alternative mixes of military manpower that can (1) meet manpower needs for specific billets without degrading performance and (2) minimize and possibly reduce the costs of manpower. This effort investigates the perception and the reality of the relationships among knowledge and skill requirements, prerequisites (e.g., education, experience), and the category of military manpower required to perform specific jobs.

RAND researchers examined opportunities to enhance or maintain personnel fill rates through alternative mixes of military manpower. One alternative mix of manpower is to increase the use of enlisted personnel, WOs, or LDOs (or civilians) in positions currently designated for officers. Our research included an analysis of how the enlisted force has evolved, a review of current manpower policy, development of a proposed framework for considering the conversion of positions held by officers to those held by others, and application of these criteria to some specific positions to provide an example of the process.

The research considers opportunities to fill funded but vacant billets with alternative manpower sources. It examines, for selected specialties, what work is currently done by officers that could be done by other categories of personnel (enlisted, WO, LDO, civilian). This work does not constitute an exhaustive list of issues or positions that the military should consider for conversion. Rather, it is an illustrative analysis of how the military may wish to approach this issue on a billet-by-billet basis.

The Evolving Enlisted Force

Though it has shrunk in recent decades, the enlisted corps has increased considerably in terms of education and experience. For example, about half of the personnel in the highest enlisted ranks now have at least an associate's degree; two decades ago, less than one in four did. Educational attainment for many senior enlisted personnel is comparable to that of WOs. In 1977, the average number of years of experience of the enlisted force was just over six years, but that experience level has risen since the inception of the all-volunteer force. The average number of years of experience of the enlisted corps is now about eight years, and it is higher still in the Air Force and the Navy. The work of enlisted and other military personnel has also changed substantially over time. Technical and craftsmen work now constitutes most military work; general military and service and supply work, traditionally the two biggest categories of military work, now constitute less than one-third of all military work. Altogether, a smaller, more educated, more experienced, more qualified, and more capable enlisted force is carrying out increasingly complex and sophisticated work for the military. This has fundamentally altered both the expectations and the capabilities of a now highly professional enlisted corps. WOs and LDOs, who come from the ranks of enlisted personnel, might offer opportunities to fill positions now held by unrestricted line officers. These very capable and very experienced personnel might therefore present an opportunity to fill vacant officer positions at, potentially, reduced manpower costs.

Review of Manpower Law and Policy

There are four sources of guidance on managing DoD manpower: United States Code (USC), the Office of the Secretary of Defense (OSD), the Chairman of the Joint Chiefs of Staff (CJCS), and the services. USC, OSD, and CJCS provide broad, overarching guidance but say very little about allocating the type of military workforce to billets. For example, USC offers broad guidance on aggregate numbers of personnel allowed in the service (e.g., how many officers are authorized in the force, the ratio of senior enlisted personnel to the rest of the enlisted force) but provides little guidance related to whether a position is to be designated a military position or what criteria should be used to designate the position for officer or enlisted performance. CJCS offers some broad and some specific guidance on the qualifications of officers and enlisted personnel assigned to the Joint Staff but leaves to the services the designation of positional requirements for officers or enlisted personnel. DoD guidance does specify criteria for positions that should be designated for performance by military personnel, but the guidance does not indicate whether the billet should be performed by an officer or an enlisted.

In sum, existing USC, OSD, and CJCS policy provides broad, overarching manpower guidance and very little guidance directing the allocation of the several types of military workforce to billets. The designation of positional requirements (i.e., whether officer or enlisted) is the domain of the services. The services therefore have flexibility to determine which positions should be filled by officers and which by enlisted personnel. The services offer much guidance on positional requirements but only limited guidance on what positions must be designated (1) as officer-only because of an officer-unique requirement or (2) as enlisted-only because of an enlisted-unique requirement. Examining alternatives to fill vacant officer positions by other manpower sources is not restricted by higher-level guidance.

If OSD takes no action and the status quo of officer vacancies remains, then ad hoc solutions will be used to address vacancies in officer manpower authorizations. This research presents an option for and a systematic approach to addressing potential solutions to manpower vacancies.

Officer Vacancies and Options to Fill Them

The project sponsor provided us with data on vacancies within the services and the COCOMs. Each service provided a listing of critical skill shortages along with inventory and requirement counts. The COCOMs also provided a list, by service, grade, number of manpower authorizations[1] and assigned personnel, and DoD occupation code, of critical manpower shortages.

To determine whether other personnel might fill officer vacancies, we considered whether it would be suitable to substitute other personnel (e.g., WOs, LDOs, enlisted personnel, civilians) or to change the role or responsibility of the billet to meet available manpower. The primary criteria we used to determine whether a position could be filled by a workforce type other than officer included whether the work was inherently military, whether it was graded for O-4s and below, and whether an officer was required to supervise other officers (i.e., for command and control or to mitigate risk). Secondary criteria for determining whether military personnel other than an officer could fill a billet included whether alternative labor sources have the requisite knowledge, skills, and abilities; whether they are used in similar jobs elsewhere; whether their use would be less costly; and whether they could fill demand.

To demonstrate how these criteria could be applied to possible positions for conversion, we selected, with the sponsor and considering current shortages and conversion possibilities, six occupations for further analysis. These were

1. unmanned aviation system (UAS) operators
2. Air Force communications and information officers
3. meteorological and oceanographic officers (weathermen)
4. Army telecommunications officers
5. Army information systems officers
6. Army signal officers.

Chapter Five provides the details of our evaluation of billets in these occupations for potential conversion, and Tables 5.1 and 5.2 summarize our application of all the criteria we used to these six occupations.

---

[1] A *manpower authorization* is a funded manpower requirement.

Recommendations

Our research indicates that DoD may have opportunities to better leverage the experience, education, and capability of the enlisted force, WOs, LDOs, and civilians in some billets currently designated for officers. To enable such substitution, DoD should promulgate guidance for designating positions for performance by enlisted personnel, WOs, LDOs, or officers. Therefore, for positions that are designated for military performance, DoD should revise Department of Defense Instruction 1100.22, *Guidance for Determining Workforce Mix,* to provide criteria and a process for the services to determine whether the military person is to be for an officer, enlisted, WO, or LDO. The research presented here provides some direction regarding—but not an exhaustive treatment of—issues to consider in developing any such guidance.

Abbreviations

CJCS	Chairman of the Joint Chiefs of Staff
COCOM	combatant command
CONUS	continental United States
CWO	Chief Warrant Officer
DoD	Department of Defense
FMS	foreign military sales
G-6	Army Chief Information Officer
KSA	knowledge, skills, and abilities
LDO	limited-duty officer
METOC	meteorological and oceanographic
MOS	military occupational specialty
NCO	noncommissioned officer
NDRI	National Defense Research Institute
NJP	nonjudicial punishment
OSD	Office of the Secretary of Defense
ROTC	Reserve Officer Training Corps
S-6	Army Communications and Information Management Staff
UAS	unmanned aviation system
UAV	unmanned aerial vehicle
USC	United States Code
WO	warrant officer

Acknowledgments

This project benefited from the input of manpower experts from each of the services who talked with us. We take responsibility for the analysis herein but acknowledge their support. We are also grateful for the support of Rich Robbins; Colonel Curt D. Smolinsky, USAF (Ret.); and Amy Parker from our sponsoring office.

We are indebted to our reviewers, John Ausink and Michael Hansen of RAND, for their thoughtful comments and improvements.

We also thank Erin-Elizabeth Johnson for editing the manuscript, Steve Oshiro for coordinating the report's publication, Sandy Petitjean for her assistance with the artwork, and Christine Galione for her support throughout the project.

The views expressed herein are our own and do not necessarily represent the policy of the Office of the Secretary of Defense.

CHAPTER ONE

Introduction

A major motivation for this research is that the services and combatant commands (COCOMs) have officer vacancies on their staffs. That is, funded officer manpower authorizations (or billets) are unfilled. Officer vacancies result when an officer with the requisite qualifications is not available to serve in a billet that requires such an officer. When this occurs, organizational performance and mission readiness can suffer. RAND researchers examined opportunities to enhance or maintain personnel fill rates through alternative mixes of military manpower. Our research included an analysis of how the enlisted force has evolved, a review of current manpower policy, development of a proposed framework for considering the conversion of positions held by officers to those held by others, and application of these criteria to some specific positions to provide an example of the process.

Several broad trends have affected the number and characteristics of U.S. military personnel in recent decades. First, the number of active-duty personnel has diminished substantially, from 2.2 million at the end of the Cold War to 1.4 million today. Second, the qualifications of military personnel continue to improve, as indicated by, for example, levels of educational attainment and performance on the Armed Forces Qualification Test. Much of the current enlisted force has received at least some college education.[1] For example, about half of the personnel in the highest enlisted ranks now have at least an associate's degree; two decades ago, less than one in four did. Educational attainment for many senior enlisted personnel is comparable to that of warrant officers (WOs). The all-volunteer force has resulted in a high aptitude, highly educated, more experienced, well compensated, and very motivated enlisted force. Third, the nature of military work continues to evolve and has become increasingly technical in recent years. The military technological revolution has changed the nature of the work that enlisted personnel perform and the training needed to perform their responsibilities. Combined, these changes have fundamentally altered the work, expectations, and capabilities of today's highly professional enlisted corps.[2] Fourth, there are vacancies in funded officer billets within the services and COCOMs. Senior enlisted personnel, WOs, limited-duty officers (LDOs), or civilians could potentially be used to fill officer billets that would otherwise remain vacant.

All this has prompted the Department of Defense (DoD) to become more interested in the cost-effective use of military manpower to meet mission demands. However, efforts to rectify the issue are complicated by the services' different approaches to meeting these demands.

[1] A recent change in compensation policy matches enlisted pay to pay in the general population for those with some college education.

[2] For additional discussion, see Kirby and Thie (1996).

The services use different categories of military manpower to meet DoD mission needs, including the categories of commissioned officer, WO, LDO, and enlisted.

Policies and practices that vary by service determine the type of military manpower (i.e., enlisted, WO, LDO or other commissioned officer) used to fill service positions. Even for what appear to be the same duties, the approaches of the services to meeting mission demands can vary. For example, the Air Force uses commissioned officers to pilot unmanned aviation systems (UASs), but the Army uses enlisted operators. To pilot rotary-wing aircraft, the Navy and the Air Force use commissioned officers who hold at least a bachelor's degree, but the Army uses WOs, many without a bachelor's degree.

Such variations can complicate the challenges that the services and the COCOMs face in filling officer positions. Alternative manpower approaches may help fill vacant billets and permit the services and the COCOMs to execute their assigned responsibilities more efficiently. In particular, expanded use of other DoD manpower resources, such as senior enlisted, WO, and LDO resources, could help bridge gaps between openings and available personnel. This report explores some alternative manpower approaches that DoD may wish to use to address gaps between vacancies and available personnel, including a reconsideration of what officer grade is actually needed in certain positions.

Objective and Approach

During our study, we examined opportunities to enhance or maintain the organizational performance of the services and the COCOMs by filling vacant funded military manpower billets through alternative mixes of military manpower, including the increased substitution of enlisted, WO, or LDO positions for officer positions.[3]

We accomplished this research through a series of tasks, conducting a descriptive, historical review and assessing how statute, policy, process, or custom led to the present state of manpower designations and how this state differs among the services.

For specific analyses, we were provided with a list of occupations with manpower vacancies as reported to the Office of the Secretary of Defense (OSD) by the services and the Joint Staff, and we also considered occupations suggested by the sponsor and our own experts. We compared the selected occupations with their closest private sector equivalents, examining both the education needed for the position and private sector practices for selecting, compensating, training, developing, and retaining personnel in the position.

We examined feasible alternatives for a different mix of military manpower than currently exists for billets in the occupations that we examined. These options were a permanent substitution of enlisted billets for some officer billets in an occupation, a grade substitution at the billet level, and the increased use of WOs and LDOs.

To assess these alternatives, we developed and used broad criteria that flow from DoD policy objectives for the use of manpower. We tailored existing DoD manpower mix criteria to assess potential alternative manpower approaches. We also developed secondary criteria that could be used as a framework for manpower categorization decisions.

We used qualitative analysis and informed judgment about the criteria to choose among the alternatives for the billets in the occupations we studied. This report presents our analysis

[3] The services might also benefit by substituting alternative manpower categories more widely.

and demonstrates a method and process that could be implemented by subject matter experts within DoD.

Finally, we synthesized the results of our research and developed associated findings and recommendations. The approach we developed for using criteria to evaluate current and future positions could be an essential element in making informed manpower decisions to best use the full capabilities of DoD's workforces.

Our work is not meant to be an exhaustive treatment of what issues DoD should consider in assessing alternative mixes of manpower. Rather, it provides a process and broad criteria that DoD may wish to consider in assessing the feasibility of such mixes for particular billets, including the qualifications of others proposed to fill them. In undertaking more-thorough consideration of alternative mixes, DoD will also need to consider, among other issues, effects on personnel management, compensation, and the cost of implementing a new mix of personnel—issues we do not consider in this report.

Organization of This Report

Chapter Two reviews how officer vacancies develop, examines the different approaches the services may take to filling these vacancies, and assesses military manpower policy and its constraints on manpower strategies. Chapter Three reviews the evolution of the enlisted force toward its present characteristics. Chapter Four posits criteria for assessing alternative manpower mixes. Chapter Five presents an assessment of these criteria for six selected occupations. Chapter Six presents conclusions and recommendations. The appendix contains a detailed review of existing DoD manpower policy.

CHAPTER TWO

Officer Vacancies, Service Manpower Approaches, and Current Manpower Policy

Officer Vacancies

The purpose of our research was to examine opportunities to fill funded officer manpower authorizations that have been left vacant by the services. We received through the project sponsor a list of requirements and inventory (i.e., authorized and assigned personnel); the results of our examination of these data appear in Chapter Five.

Officer vacancies, and the challenges resulting from them, occur when an officer with the requisite qualifications is not available to serve in a billet that requires such an officer. When this occurs, the mission readiness of the unit with the vacant position can suffer. Even when officers are not available to perform these assignments, the work remains and must still be performed.

Alternative manpower sources could be used to fill vacant officer billets. Other categories of manpower might support and efficiently execute some responsibilities that the services and the COCOMs currently designate for officers. Expanded use of the enlisted, WO, and LDO (and civilian) workforces could help bridge gaps between available officers and positions nominally requiring officers.

Service Manpower Approaches

The services have different manpower and personnel systems, practices, and policies. They sometimes arrive at different manpower solutions to meet a very similar need. For example, the Air Force currently uses only rated pilots to fly UASs, but the Army uses noncommissioned officers (NCOs) in this role. Although the UAS airframes flown by the Air Force and the Army were initially different, the two services now procure similar airframes that fly similar missions. Nonetheless, there still exists a discrepancy between the two services in terms of the competencies (knowledge, skills, and abilities [KSAs]) and the prerequisites (e.g., education, experience) that each has identified as needed to effectively perform as a UAS operator.

Although the Army and the Air Force have taken different approaches to choosing the type of personnel permitted to pilot UASs, the rationale for the approach used may relate to the missions or locations in which each service's UASs fly. For example, the Air Force aircraft may fly longer or higher-altitude missions that could cause the UAS to venture into controlled airspace, which requires a rated pilot at the controls. Army UASs fly tactical missions in support of ground units, which does not typically take the aircraft into controlled airspace. Therefore, there is a specific reason why the Air Force assigns only rated officers as UAS pilots and why the Army does not. This example shows that different services apply different criteria to what appear to be similar jobs, and it demonstrates how they come to different conclusions about the appropriate workforce type.[1]

The remainder of this section supplies a general assessment of the processes and policies used by the services to meet the demand for personnel with the supply of personnel, beginning with the Army.

The Army designates positions by considering the organizational setting of the billet, the positional authority and responsibility of the billet holder, the criticality of the job to the organization's mission, and the skills and knowledge required.

The Navy, in designating positions for officers or enlisted personnel, considers the engineering design of a system, mission statements in the unit's required operational capability and projected operational environment, command and control and organizational functions, specialized skills needed, and how similar units have been manned in the past. Enlisted positions are manned based on occupational standards and Navy enlisted classifications.

The Air Force manpower requirements are established by individual major commands and then adjudicated by the Air Staff. The Air Staff uses three approaches in evaluating positional assignments: comparing job descriptions, classifying results, and understanding and evaluating the differences in the type of work done and time spent performing various work activities. The Air Force uses a quantitative approach that measures the types of tasks that are done and the percentage of time a billet holder spends performing each task. This occupational analysis approach both identifies and quantifies job requirements, and it is used to specify training demands, personnel selection, promotions testing, and other functions.

Current Manpower Policy

The roles that officers and enlisted personnel can currently fill are determined by four sources of guidance on managing DoD manpower: United States Code (USC), OSD, the Chairman of the Joint Chiefs of Staff (CJCS), and the services. We review these four sources in detail in the appendix but summarize their implications for alternative manpower strategies in this chapter.

We examined the guidance from these four sources from several perspectives. We began with a review of the general guidance (i.e., the broad direction of law and policy on DoD utilization of manpower) and then examined overall application of the guidance to military personnel. We next focused our review to determine how the guidance affected the utilization of officers and enlisted personnel. Finally, we examined both the use of DoD civilians and the policies that govern their use.

We conducted this review to determine how statute, policy, and process led to the present state of determining manpower requirements and how this state might differ among the services. The objective of our review was to capture the specific directives requiring that military personnel be assigned to positions or billets. We also sought to determine, for positions designated as military, why officers or enlisted personnel might be assigned to them.

---

[1] See Chisholm (2007) and Murnan (2008) for a more detailed discussion.

Limited Policy on Military Manpower Mix

USC and DoD and CJCS directives provide only broad, overarching guidance on manpower. They say very little about allocating military workforces to billets. USC offers broad guidance on raw numbers, including the number of authorized officers and the ratio of senior enlisted personnel to other enlisted personnel. However, it provides little guidance on the categorization of billets for officers or enlisted personnel.

In Department of Defense Instruction 1100.22, *Guidance for Determining Workforce Mix*, DoD provides overarching guidance to the military departments for determining the workforce mix. This guidance, however, only provides criteria for determining whether a position should be designated as military or civilian. When the guidance requires that a position be designated for military performance, it does not address how to determine whether an officer or an enlisted should fill the position.

CJCS directives offer some broad and some specific guidance on the qualifications of officers and enlisted personnel assigned to the Joint Staff. This includes specific circumstances requiring officers to serve in designated assignments. For example, CJCS guidance requires officers with specific experience to serve in joint positions.

The designation of positional requirements for officers or enlisted personnel, however, is the domain of the services and is not specified in DoD policy. The services provide specific internal guidance on the KSAs needed to effectively meet positional requirements, but their approaches to setting these requirements—much less determining whether officers, enlisted personnel, or even civilians can meet them—vary.

Overall, DoD directives primarily focus on determining whether work must be performed by military personnel or can be performed by civilians. There is little guidance on whether specific types of military work should be performed by officers or enlisted personnel. The services offer much guidance on positional requirements but only limited guidance on why positions may be designated for officers or for enlisted personnel.

Positions for Military Personnel

As noted in the previous section, DoD guidance does specify criteria for designating positions for performance by military personnel. The guidance states that a position shall be designated as military when required

- for reasons of law
- for command and control of crisis situations
- for combat readiness
- for esprit de corps
- because of unusual working conditions not conducive to civilian employment
- to ensure that the military-unique knowledge and skills required for successful performance of the duties are available.

Clearly, some of these requirements will lead to designation of an officer for a position. For example, officers traditionally serve in positions involving command and control of crisis situations and assuring the combat readiness of assigned forces.

DoD guidance further states that there must be sufficient military manpower to provide a rotation of personnel assigned outside the United States and for sea-to-shore rotation. Designation of positions as military must also consider career opportunities for military personnel, including the development of combat-related skills and other military competencies. USC also requires a few positions be held by an officer, including those that involve exercising nonjudicial punishment (NJP) authority over subordinates.

Service staffing guidance provides some detail regarding when different categories of personnel should fill specific positions, but the guidance is often broad and unspecific. For example, Navy guidance on manpower requirements stipulates that

- An officer (except a chief warrant officer [CWO][2]) is required for making managerial, operational, or professional (as applied to officer specialty category) decisions. Yet, some managerial positions may also be classified for enlisted personnel at appropriate managerial levels.
- An LDO[3] is acceptable if performance of duties is limited to specific occupational fields, requires authority and responsibility greater than normally expected of a CWO, requires managerial skills, and is outside the normal development pattern for unrestricted line and restricted line officers.
- A CWO is acceptable if performance of duties is limited in scope (in comparison with other officer categories), is technically oriented (i.e., requires relevant experience and specialized training), and is repetitive in nature.[4]

Navy guidance also states that, when classifying a position as appropriate for an LDO or CWO, the LDO/CWO/E-9/E-8 functional relationship mix must be taken into account to preclude excessive layering of officer and enlisted technical/managerial talent. For example, the Navy seeks to avoid assigning an E-8 or E-9 as subordinate to a CWO or assigning a CWO to an LDO below an O-3 rank.

LDOs, CWOs, and senior enlisted personnel possess overlapping KSAs and experience. As a result, senior enlisted personnel are considered for staffing at "appropriate" managerial levels.

Like the Navy, the Army, the Air Force, and the Marine Corps have independently developed manpower assignment policies. The appendix to this report contains a list and detailed review of existing manpower policy, repeating sections of USC, DoD Directives, and service directives and guidance on the designation of the type of personnel appropriate for manpower billets. Our goal in reviewing these references was to determine whether existing criteria or guidance would preclude enlisted personnel, WOs, or LDOs from filling vacant officer positions.

---

[2] A CWO is an officer ranked above the most senior enlisted and officer cadets and candidates but below the officer grade of O-1. WOs possess a high degree of specialization in a particular field in contrast to the more general assignment pattern of other commissioned officers. WOs command aircraft, maritime vessels, special units, and task-organized operational elements. WOs are technical experts who operate, maintain, administer, and manage equipment, support activities, and technical systems.

[3] An LDO is an officer in the Navy or Marine Corps who was selected for commissioning based on his or her skill and expertise. An LDO is not required to have a bachelor's degree. LDOs have technical knowledge and expertise and are seasoned leaders. An LDO is a commissioned officer designated for limited duty.

[4] See Navy Manpower Analysis Center (2005).

Each service follows its own method in determining what types of personnel are needed to fill manpower assignments, but the services' approaches to assigning officers and enlisted personnel do not differ dramatically. We found that, although there clearly are billets that must be filled by officers (e.g., billets for officers in command, billets involving NJP authority, billets for Joint Staff positions), the requirement for officers to fill manpower assignments is dictated by service policy, not by a higher authority. DoD and higher authorities provide broad—not prescriptive—guidance to the services regarding the determination of the type of personnel to fill manpower positions. Therefore, the services have the discretion and latitude to change the type of manpower that they deem necessary to fill manpower positions.

CHAPTER THREE

An Evolving Enlisted Force

Today's enlisted workforce is the result of countless short-term responses to a personnel system developed in the 1950s and then shaped by conscription in the 1960s and early 1970s; the end of conscription in 1973; significant compensation and benefit changes in the 1980s; and significant changes in size, organization, and technology throughout the years.

Of perhaps the greatest significance to considerations of converting some positions from those held by officers to those held by others is that the quality of the enlisted corps has changed over time. The all-volunteer force has resulted in a high aptitude, well educated, more experienced, well compensated, and very motivated force that performs at a high level. The overwhelming majority of military recruits today are high school graduates who also scored well on the Armed Forces Qualification Test. In 2003, for example, 95 percent of new recruits were high school graduates, and 72 percent scored above average on the enlistment aptitude test. By comparison, less than 80 percent of American youth have a high school diploma (and half, by definition, score below the median on the aptitude test).[1] In the past 20 years, educational attainment has increased, particularly for senior enlisted personnel, with many now having associate, baccalaureate, and advanced degrees. Since the early 1980s, the average enlisted experience in terms of number of years of service has also increased.

DoD has made many efforts to keep pay competitive with the private sector, resulting in high retention rates for today's force. Because it is highly motivated and well compensated, there are great demands placed on the enlisted force, and performance expectations are high.

In this chapter, we first review the changing environments that set the demand for alternative manpower strategies. We then examine the transformation of the enlisted force, to include increased educational attainment, increased experience, and the structure and nature of work performed by this force. We broadly discuss how an enlisted person ascends to become a WO or LDO and address the type of work that these personnel perform. We also discuss the civilian labor force and provide examples of enlisted personnel, WOs, and LDOs doing "officer work."

Changing Environments and the Need for Alternative Manpower Strategies

Two broad groups of considerations, depicted in Figure 3.1 and discussed in greater detail later in this section, may help shape alternative manning strategies. The first group is legacy manning issues, including an officer-heavy force and the expense associated with developing and

[1] Rostker (2006).

Figure 3.1
Environmental Factors That May Shape Alternative Manpower Strategies

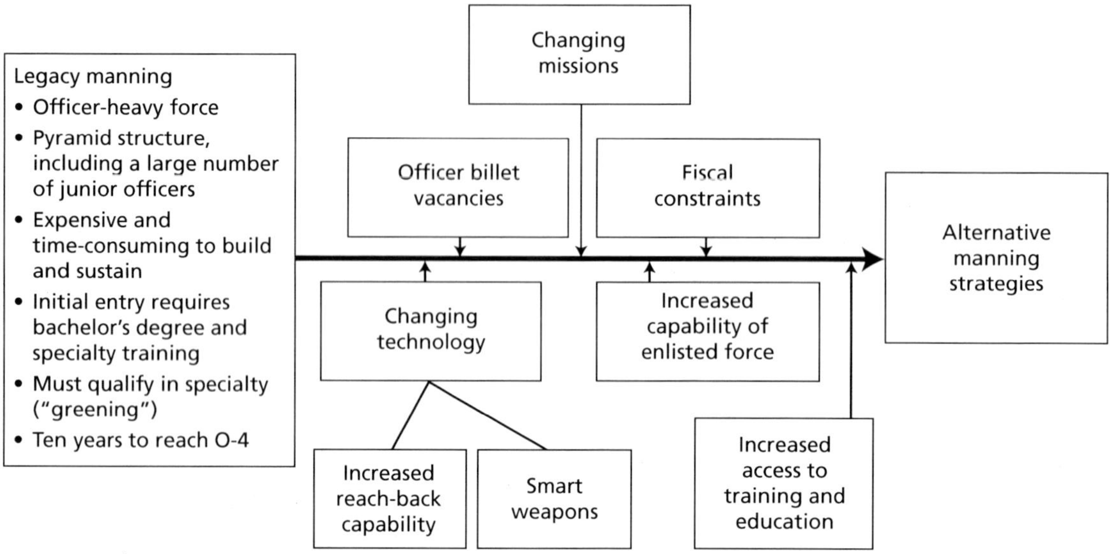

RAND *TR881-3.1*

maintaining such a force. The second group comprises the following dynamic environmental characteristics: changing missions, vacant billets, fiscal constraints, changing technology, the increased capability of the enlisted force, and increased access to training and education.

Legacy Manpower

The structure of the officer force resembles a pyramid, with a large number of junior officers entering from the bottom, a smaller number of midgrade officers, and relatively few senior officers at the top. In this closed, time-based system, developing officers takes a long time. At entry, junior officers typically have a four-year baccalaureate degree and up to two years of occupational specialty training (e.g., for a pilot).[2] After receiving specialty training, an officer must still take time to qualify in his or her specialty, including time to master the basics of the occupation. Officers must maintain both generalist and specialist qualifications to advance. Altogether, it takes about ten years of active service (plus precommissioning education and training) for officers to reach the grade of O-4, at which point they have achieved the level of knowledge and experience to be considered a specialist and to serve in service staff positions. The rigors and costs of this process and the need to make sure that officers are assigned where they are most needed and not just where policy may dictate are among the principal reasons for considering alternative manning strategies.

Environmental Factors

Changing missions also suggest the need for a more dynamic approach to addressing manpower needs. The U.S. military no longer faces an opposing strategic force. Rather, its missions today include countering insurgencies, terrorism, and piracy and involve repeated deployments in support of Operation Iraqi Freedom and Operation Enduring Freedom. The knowledge and

---

[2] If a Reserve Officer Training Corps (ROTC) center or service academy is the source of the officer's commission, then that adds to the timeline of the developmental pipeline.

skills needed to undertake these missions are evolving. Experience in the field may be a better prerequisite for mission success than knowledge obtained through long education and courses of instruction.

Fiscal constraints may cause DoD planners to examine the most-effective use of military manpower while minimizing costs. It is appropriate to consider using less-costly military personnel (e.g., enlisted personnel and, potentially, WOs, LDOs, and civilians) rather than officers if the former can meet organizational and mission requirements.[3]

Changing and increasing technological capabilities can also shape alternative manpower strategies. Military forces today have increased reach-back capabilities. Through high-speed communications and technological advances, forward-deployed forces in the field and in the fleet can tap into knowledge and expertise elsewhere to obtain real-time support for meeting mission demands. "Smart" weapons are another technological capability that is a force multiplier. Manpower approaches and tactics involving these weapons must keep pace with changing technology.

Increased access to training and education can continue to improve the qualifications of the force available to meet military missions. Online and other computer-based training provides users with greater access to training and education and with greater ability to increase their knowledge, assume new positions, and meet new demands.[4]

Finally, DoD must consider what level of efficiency, capability, and performance could result from alternative force structures. In particular, are there alternative manpower approaches, such as increased use of senior enlisted personnel in officer billets, that can effectively meet mission needs at reduced cost?

All in all, U.S. forces operate in an increasingly dynamic environment. Accordingly, to meet mission challenges, they may need a more dynamic manpower approach, including alternative manpower strategies for filling positions that may be difficult to fill because of imbalances between the demand for and supply of commissioned officers.

Enlisted and Warrant Officer/Limited-Duty Officer Educational Attainment

Many first-term personnel now obtain some college education.[5] Between 1988 and 2008, senior enlisted (E-7, E-8, and E-9) educational attainment steadily increased and, in 2008, most E-9s, nearly half of E-8s, and about one-third of E-7s held at least an associate's degree (see Figure 3.2). This represents a significant increase over levels from 20 years ago.

Like many senior enlisted personnel, many WOs also have high levels of educational attainment (see Figure 3.3). This increased level of education may indicate potential for more fully leveraging the WO workforce. Educational attainment is also high for commissioned

[3] A number of studies have looked at the cost of various workforces in proposing military-to-civilian or other manpower conversions. This study considers those results as one criterion for reviewing billets but is focused on other aspects of the feasibility of substituting other types of manpower for vacant officer billets.

[4] See Cavallaro (2008) for a discussion of how military training and education for enlisted personnel are changing.

[5] Department of Defense, Under Secretary of Defense for Personnel and Readiness (2002).

Figure 3.2
Enlisted Educational Attainment, Paygrades E-1 Through E-9, 1988, 1998, and 2008

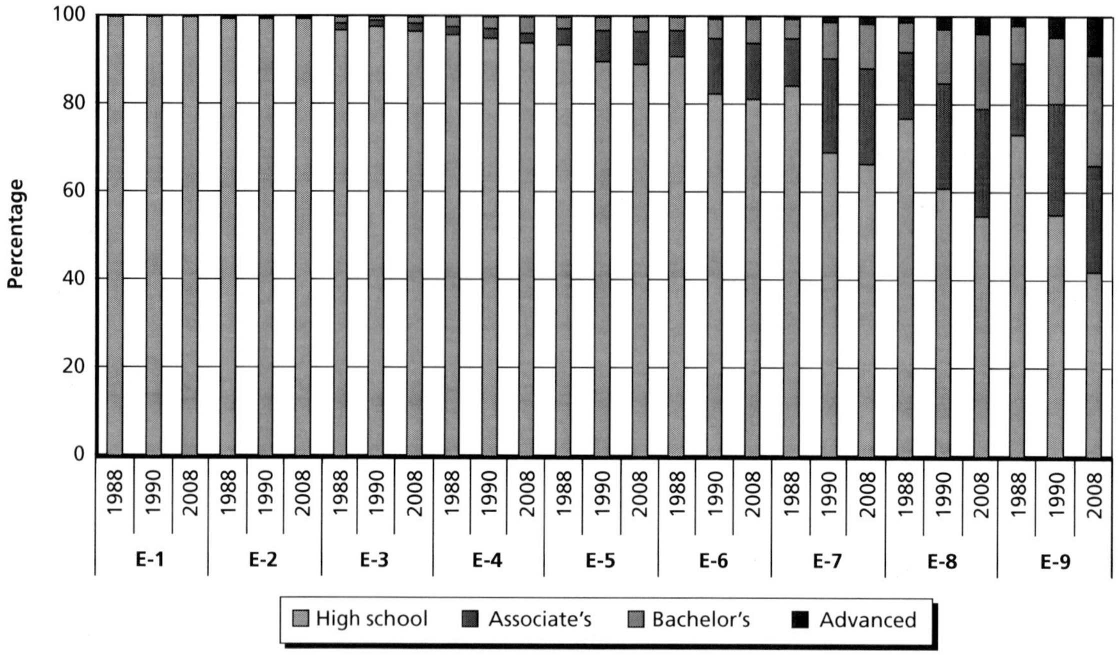

SOURCE: Defense Manpower Data Center information supplied in 2009.
RAND *TR881-3.2*

Figure 3.3
Educational Attainment, Paygrades E-1 Through O-10, 2008

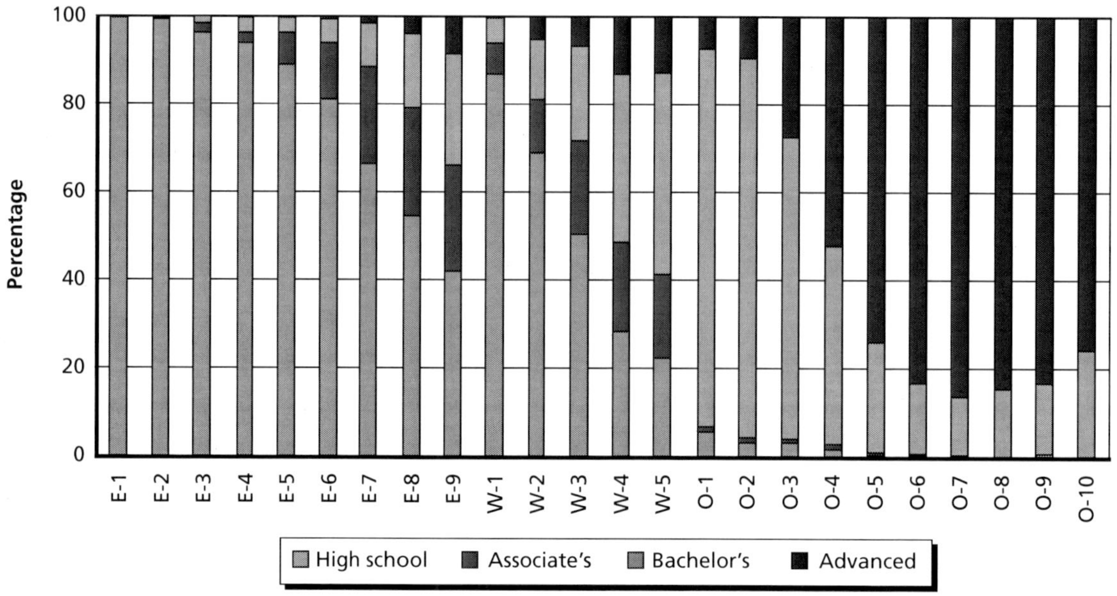

SOURCE: Defense Manpower Data Center information as of 2008.
RAND *TR881-3.3*

officers: As of 2008, about 80 percent of those with a grade of O-5 or higher hold an advanced degree.[6]

Enlisted Experience

The average number of years of military experience of the enlisted workforce has increased since the advent of the all-volunteer force (Figure 3.4). Experience levels significantly increased between 1980 and 1995, remained steady between 1995 and 2006, and slightly decreased in 2007. In 1977, the average number of years of experience of the enlisted force was just over six years. The experience level of the enlisted force rose to eight years and remained at that level through 2005 before dropping slightly. DoD now enjoys a military workforce with an average of just less than eight years of service. The increased experience level has helped reduce the number of new accessions needed to man the force.

Just as the experience level of the enlisted force has increased across DoD, so too has it increased within each service (Figure 3.5). Average experience is highest in the Air Force and next-highest in the Navy, where changes in average experience have roughly paralleled those in the Air Force. Average experience in the Army and the Marine Corps has decreased in recent

Figure 3.4
Enlisted Force Experience, 1953–2007

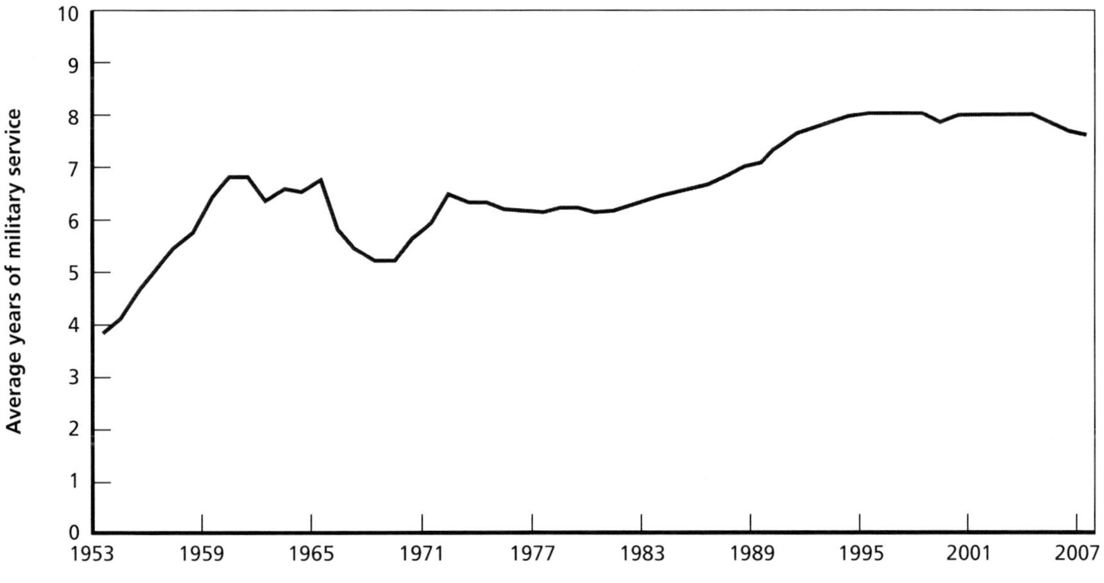

SOURCE: DoD information supplied in 2009.
RAND *TR881-3.4*

---

[6] Gottschalk and Hansen (2003) use the college compensation premium within an occupation to identify whether a job is "college" or "non-college." However, the military compensates largely on the basis of attained rank and longevity. There are no premiums for college or graduate educational attainment. Traditionally, officer jobs have been considered college jobs and enlisted jobs have not. In fact, law for reserve officers and service policy for active-duty officers require that all commissioned officers have a baccalaureate degree at entry or by the grade of O-3, so, by fiat, officer jobs are college jobs. Even absent the college premium to sort jobs in the military, one could argue either that some enlisted jobs have become college jobs because of skill-based technological change or that certain enlisted personnel with baccalaureate or higher educations could be used in traditional officer college jobs.

Figure 3.5
Enlisted Force Experience, by Service, 1953–2007

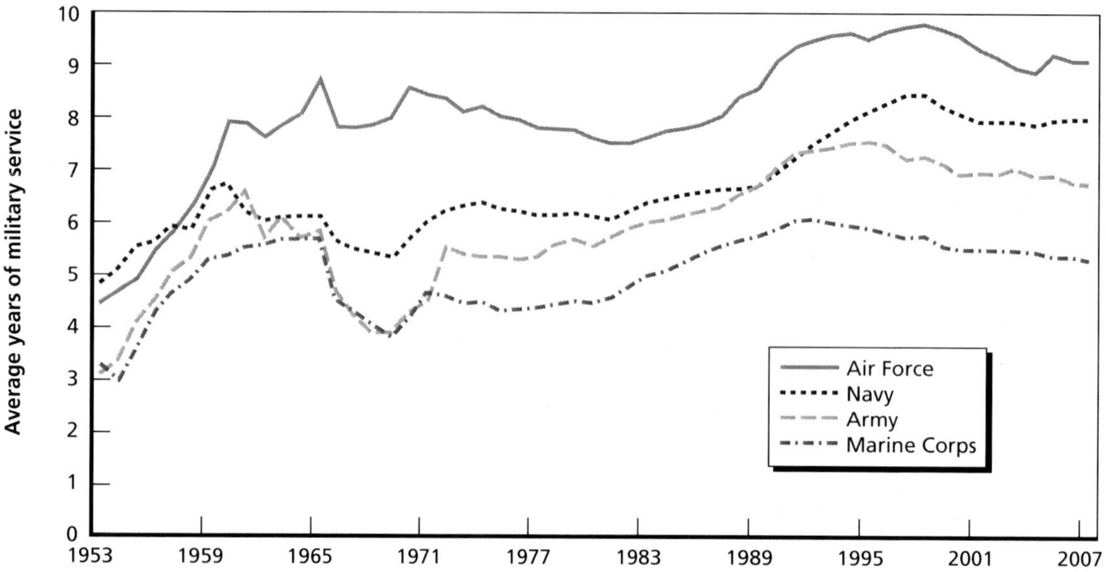

SOURCE: DoD information supplied in 2009.
RAND *TR881-3.5*

years as those components have added personnel. Enlisted personnel in the Marine Corps, which has a relatively junior force by design, have the lowest average level of experience.

The growing experience and education of enlisted personnel raises questions about how that force might be better used in certain positions. Could increased experience and education support increased assumption of leadership responsibilities? Do work efficiencies result from a more senior force? What benefits are derived from a more senior, more experienced force?

The Enlisted-to-Officer Ratio

The ratio of enlisted personnel to officers has changed over time. After sharply declining as the Vietnam War ended, the ratio of enlisted personnel to officers increased to 6.4:1.0 in 1977 (Figure 3.6). The ratio has decreased steadily since, standing today at a little more than 5:1. Ratios vary by service, ranging from approximately 9:1 in the Marine Corps to 5:1 in the Army and the Navy to 4:1 in the Air Force. Low ratios of enlisted personnel to officers reflect an officer-heavy—and more expensive—force.

The Changing Nature of Military Work

The nature of U.S. military work has changed over time. In the late 19th century, the overwhelming majority was general military[7] work; even through World War II, such work com-

[7] The categories discussed in this section are aggregates based on the DoD Occupational Coding System. General military work is DoD Occupational Code 1, infantry and seamanship. All military services have some personnel in this category, but the Army and the Marine Corps have the most. Technical work is the aggregate of the following DoD occupational

Figure 3.6
The Enlisted-to-Officer Ratio, 1955–2007

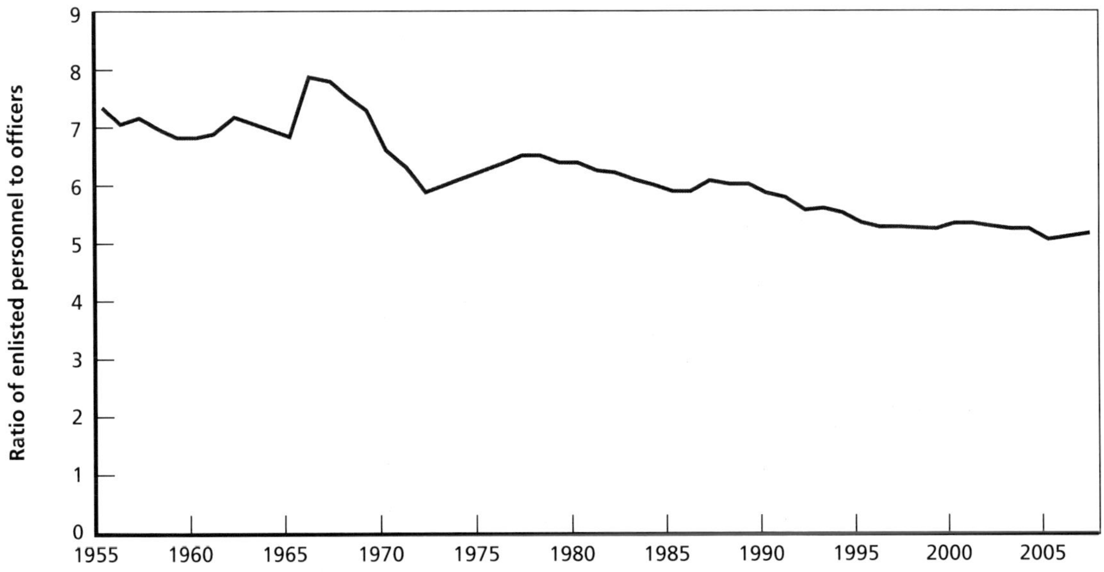

SOURCE: DoD information supplied in 2009.
RAND *TR881-3.6*

prised the single largest category of military work (see Figure 3.7). During World War I, service and supply work was the second largest category of military work, but such work has decreased since then. Most recently, there has been a transfer of enlisted occupations away from services and clerical work to other areas, particularly technical work. As the military has become more technically oriented, more trained technicians have been needed to operate, maintain, and troubleshoot electronic equipment. The increase in DoD's technical workforce as a percentage of military work has been substantial: As of 2005, the technical workforce accounted for approximately 30 percent of DoD's total workforce.

The evolution of the all-volunteer force may be presenting the military with new ways to meet the demands of its increasingly complex work. The KSAs and prerequisites (e.g., education, experience) needed to effectively perform military jobs must take into account the expanded capabilities of the enlisted force. This suggests that past practices permitting only officers to fill some positions should be reviewed. Requiring that an officer with a bachelor's degree fill certain positions may be an outdated practice that fails to recognize both the experience and capabilities of the enlisted force and the value of having that force undertake new and evolving positions.

Warrant Officers, Limited-Duty Officers, and Civilians

WOs and LDOs begin their careers as enlisted personnel. The services provide senior-graded personnel with a path to become officers through WO and LDO programs. Senior enlisted

codes: communications and intelligence, electronic, medical and dental, and other technical. Clerical work is the support occupational code. The craftsmen category includes the electrical and mechanical occupational codes. Service and supply is the service occupational code.

Figure 3.7
The Nature of Military Work, 1865–2005

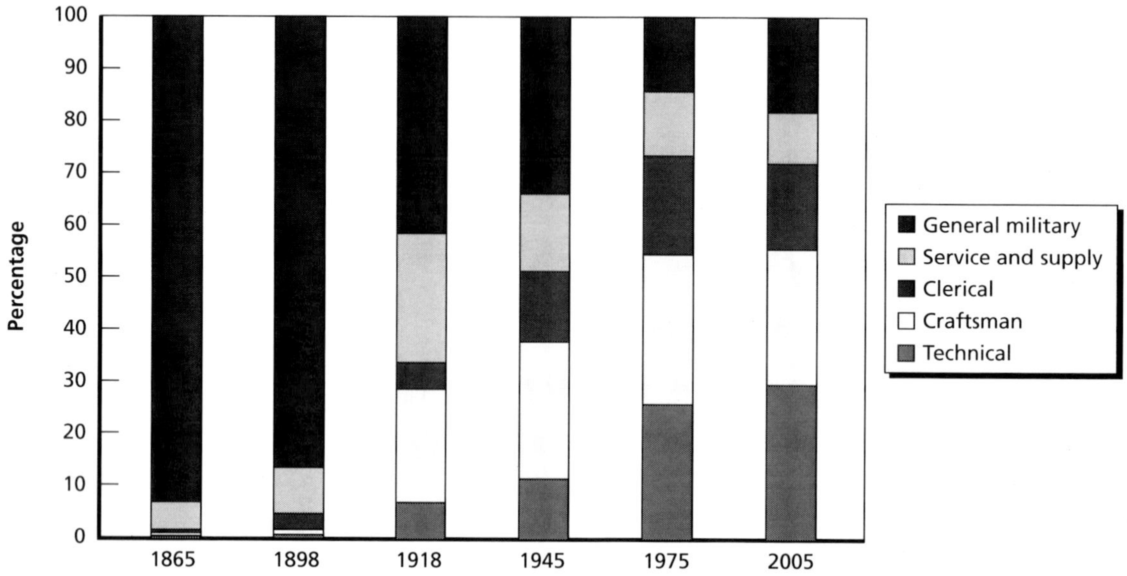

SOURCE: Eitelberg (1988) and Defense Manpower Data Center information supplied in 2009.
RAND *TR881-3.7*

personnel (E-7s through E-9s) can apply for WO and LDO programs and, if selected, serve as technical experts and managers in their specialized community. Because WOs and LDOs begin their careers as enlisted personnel before gaining a commission, they bridge the gap between the ranks of enlisted personnel and officers.

Warrant Officers

WOs are technical leaders and specialists in the military. They serve in the Army, the Navy, and the Marine Corps. They perform technical duties relating to skills acquired through previous enlisted service and training.

In the Army, in contrast to the more general assignment pattern of other commissioned officers, WOs are highly specialized in a particular field. They command aircraft, maritime vessels, special units, and task-organized operational elements, and they operate, maintain, administer, and manage the Army's equipment, support activities, and technical systems. As WOs begin to function at the higher levels, they become "systems-of-systems" experts rather than experts in specific equipment.[8]

In the Navy, CWOs are technical specialists who perform duties requiring extensive knowledge of and skills related to a specific occupational field. The WO program is open to enlisted personnel who have completed at least 12 years of service and have earned the rank of chief petty officer or higher.

[8] Derived from FM 6-22.

In the Marine Corps, a WO is a technical officer specialist who performs duties that require extensive knowledge of and training and experience in the employment of particular capabilities that are beyond the duties and responsibilities of senior NCOs.[9]

Limited-Duty Officers

LDOs serve in the Navy and the Marine Corps. Like WOs, they are not required to have a bachelor's degree; however, they have technical knowledge and are seasoned leaders. LDOs perform work that is similar to that performed by WOs, and differences between LDOs and WOs are subtle. The authority of LDOs is generally limited (or restricted) in that LDOs cannot command a unit. However, in some cases, LDOs may command activities whose primary function corresponds to their specialty. In the Navy, LDOs are technically oriented officers who perform duties in specific occupational fields and require strong managerial skills.

In the Marine Corps, LDOs are technical officer specialists who perform duties that require extensive knowledge of, training in, and experience with the employment of particular capabilities that are beyond the duties and responsibilities of a WO and that cannot be met by an unrestricted officer. LDOs in the Marine Corps serve in the grades of captain (O-3), major (O-4), and lieutenant colonel (O-5). Marine Corps LDOs are selected from the WO ranks, and WOs must have reached eight years of service as a WO to be considered for selection as an LDO.

In the Navy, the LDO and CWO programs provide commissioning opportunities to qualified senior enlisted personnel. Chief petty officers (E-7s through E-9s), E-6 personnel who are selection-board eligible for E-7, and CWOs (applying for lieutenant [junior grade]) may qualify for these programs. The LDO and CWO programs are open to both active-duty and inactive-duty personnel. A baccalaureate degree is not required; however, it is strongly encouraged.[10]

WOs and LDOs could potentially perform work that is currently done by (more-senior) officers without degrading performance, as could civilians. The financial compensation that they receive depends on their occupation and experience level. Therefore, marginal differences in manpower cost savings might be realized when WOs, LDOs, and civilians are substituted for officers, as the former group are generally more highly compensated than enlisted personnel. WOs and LDOs can and do perform work that was previously done by more-senior officers, as shown later in this chapter.

Civilians

The use of civilian personnel (government and contractor) to fill vacant officer positions is another option worthy of consideration. Indeed, the Secretary of Defense took such action in the military-to-civilian conversion program in fiscal year 2006. Before converting a military billet to civilian position, however, the billet must not be deemed "military essential."

Civilians can and do possess the KSAs to perform in some officer assignments. However, the cost-effectiveness of such a conversion can be difficult to ascertain. Cost-effectiveness is one priority among many for performing military-to-civilian conversions, and, although it must be

---

[9] Derived from SECNAVINST 1412.9B.

[10] Derived from OPNAVINST 1420.1A.

weighed against other priorities, such as the need for a rotation base and career management considerations, there is pressure to achieve a more cost-effective workforce.

Examples of Enlisted Personnel and Warrant Officers Performing Work Traditionally Done by Line Officers

The services have increasingly used enlisted personnel, WOs, and LDOs in roles traditionally filled by line officers. With the exception of the Air Force, all of the services use WOs for technical and technical specialist jobs. The Navy and the Marine Corps also use LDOs for these positions. WOs and LDOs do not need a bachelor's degree to perform their duties but are technical experts in their field.

In an earlier section, we addressed some of the many responsibilities that Army WOs are tasked to perform. Here, we note that the Army uses WOs to fly virtually all types of aircraft, including fixed-wing and armed rotary aircraft (i.e., Apache helicopters). Both the Army and the Navy use enlisted personnel to pilot UASs.

The Navy has also provided opportunities to enlisted personnel in its Smartship minimal manning program. In minimally manned ships, a decreasing number of personnel have taken on increasing responsibilities. Other examples of Navy enlisted personnel now performing jobs traditionally done by officers include the following:

- In 2009, the Navy introduced a program allowing WOs to be aviators and flight officers aboard fixed-wing P-3 Orion and H-60 Blackhawk aircraft. A newly created aviation program for highly qualified sailors (E-5s through E-7s) allows selected personnel to become CWOs, to undergo flight training, and to serve in cockpits as pilots and as naval flight officers. This program creates flying specialists unencumbered by the traditional career paths of the unrestricted line community.[11]
- E-5s and higher-level enlisted personnel are serving in the position of engineering officer of the watch, with responsibility for the safe and effective operation of the engineering plant on an underway ship.
- Chief petty officers (E-7s) are standing watch as officers of the deck under way, with responsibility for the safety and security of the ship. On the Navy's new Littoral Combat Ship, which has a crew of just 40, enlisted crewmembers perform many roles, and the traditional responsibilities of officers and enlisted personnel are blurred in other ways.

---

[11] See NAVADMIN 067/09.

CHAPTER FOUR

Criteria to Evaluate Alternative Manpower Options

In conducting our analysis, we sought to list factors that describe senior officer and senior enlisted military personnel. In Table 4.1, we provide a general description of and list the major occupational groups that comprise senior officers and senior enlisted personnel. We then compare the general demographics of each group, including grade, average number of years of service required to attain the grade, fiscal year 2009 composite pay costs to DoD, and education level. We also include a general description of the training and duties associated with each grade.

Senior enlisted personnel have 10–22 years of service experience; an O-4 has at least ten years of service. As we noted in an earlier chapter, many senior enlisted personnel have associate's and bachelor's degrees, while junior officers tend to have bachelor's degree or higher. Senior enlisted personnel have advanced technical training, while junior officers attain specialized technical training.

Senior enlisted personnel develop deep technical knowledge and leadership experience, while many junior officers are developing and gaining experience in their positions. The nature of work for junior officers requires them to be accountable decisionmakers, whereas senior enlisted personnel are managers and technicians.

The scope of work varies by billet. Junior officers generally integrate the unit's capabilities into a larger scheme and perform mission planning. Senior enlisted personnel employ the unit's capabilities and execute orders. Junior officers and senior enlisted personnel manage a small number of subordinates through indirect and direct supervision.

Both junior officers and senior enlisted personnel are subordinate to command authority and are responsible for the performance of assigned personnel. Junior officers have authority over officers and assigned enlisted personnel, whereas senior enlisted personnel have positional and grade authority over subordinates.

The training, experience, nature and scope of work, span of control, and duties and responsibilities of senior officers are very different from those of senior enlisted personnel. Senior officers (O-5s and above) possess advanced education, advanced training, and deep experience; are accountable decisionmakers; have enterprise knowledge; provide strategic direction; and exercise a large span of control and a high level of responsibility. Junior officers are developing these attributes and characteristics, but they have limited experience, limited management education, and limited functional or technical knowledge. Senior enlisted personnel, WOs, and LDOs have experience, management ability, and technical knowledge. They are used in operational and technical positions.

Our assessment and comparison of factors pertaining to junior officers (O-4s and below) and senior enlisted personnel indicate that, in terms of general training, education, and duties, senior enlisted personnel have more in common with junior officers than with senior officers.

Table 4.1
General Factors Describing Senior Military Officers and Senior Enlisted Personnel

	Officer			Enlisted		
General description	Officers are the leaders of the military, supervising and managing activities in every occupational specialty.			Enlisted personnel carry out the fundamental operations of the military in combat, administration, construction, engineering, health care, human services, and other areas.		
Major occupational groups	Combat specialty; engineering, science, and technical; executive, administrative, and managerial; health care; human resource development; media and public affairs; protective service; support services; and transportation.			Administrative; combat specialty; construction; electronic and electrical equipment repair; engineering; science; and technical; health care; human resources development; machine operator and production; media and public affairs; protective service; support service; transportation and material handling; and vehicle and machinery mechanics.		
General demographics						
Grade	O-6	O-5	O-4	E-9	E-8	E-7
Years of service	≥ 22	≥ 16	≥ 10	≥ 22	≥ 16	≥ 10
Cost[a]	$206,176	$174,991	$160,281	$129,962	$110,557	$99,340
Education level	Bachelor's degree; funded master's degree program; advanced degree	Bachelor's degree; funded master's degree program	Bachelor's degree	Associate's degree; some bachelor's education	Associate's degree or more	Some college
General description of training and duties						
Training	Advanced specialty training; highly specialized, deep officer education and experience	Advanced specialty training; highly specialized, deep officer education and experience	Advanced specialty training; specialized officer education and gaining experience	Advanced technical training; specialized enlisted education, training, and experience	Advanced technical training; specialized enlisted education, training, and experience	Advanced technical training; specialized enlisted education, training, and gaining experience
Experience	Requires deep experience in officer positions	Requires deep experience in officer positions	Position is used to gain developmental experience	Requires deep technical and/or leadership experience	Requires deep technical and/or leadership experience	Requires technical and leadership experience
Nature of work	Executive/commanding officer; accountable decisionmaker	Organizational leader/commanding officer/staff officer; accountable decisionmaker	Executive officer/ staff officer/ manager; accountable decisionmaker	Senior enlisted leader/ manager	Senior enlisted leader/ technician/first-level supervisor	Technician/first-level supervisor

	Officer			Enlisted		
Scope of work	Warfighter; warfighting policymaker; strategic/operational/tactical; responsible for employing capabilities, integrating unit capabilities (could be those of self and the unit) into the larger scheme, integrating external capabilities into the unit, and mission planning	Warfighter; operational/ tactical; responsible for employing capabilities, integrating unit capabilities (could be those of self and the unit) into the larger scheme, integrating external capabilities into the unit, and mission planning	Warfighter; operational/tactical; responsible for employing capabilities, integrating unit capabilities (could be those of self and the unit) into the larger scheme, and mission planning	Warfighter; tactical; creates an effect by employing unit capabilities; executes orders	Warfighter; tactical; creates an effect by employing unit capabilities; executes orders	Warfighter; tactical; creates an effect by employing unit capabilities; executes orders
Span of control	Can manage/command a large number of units or people; conducts indirect and direct supervision; manages the work of organizations through subordinate leaders	Can manage/command a medium number of units or people; conducts indirect and direct supervision; manages the work of organizations through subordinate leaders	Manages a small number of people; conducts indirect and direct supervision	Manages a medium number of people; conducts indirect and direct supervision	Manages a medium or small number of enlisted personnel; conducts indirect and direct supervision of individuals producing direct and measurable output	Manages a small number of personnel; conducts indirect and direct supervision of individuals producing direct and measurable output
Level of responsibility	Major command; can have policy implications; represents the unit to external organizations at different echelons based on own judgment	Battalion/unit command; represents the unit to external organizations at different echelons based on own judgment	Subordinate to command authority; responsible for performance of duties of assigned personnel	Command senior enlisted leader; must conform to existing policy, as defined by an officer; coordinates through hierarchy	Department/division chief; must conform to existing policy, as defined by an officer; coordinates through hierarchy	Division chief; must conform to existing policy, as defined by an officer; coordinates through hierarchy
Command/ NJP authority	Exercises command and NJP authority over units or subordinates	Exercises command and NJP authority over subordinates	Has direct authority over subordinate junior officers and enlisted	Positional/grade authority over subordinates	Positional/grade authority over subordinates	Positional/grade authority over subordinates
Subordinates supervised	Officers and enlisted	Officers and enlisted	Officers and enlisted	Senior and junior enlisted	Senior and junior enlisted	Junior enlisted

SOURCES: U.S. Bureau of Labor Statistics (2008–2009), for general descriptions of duties and occupational groups; Defense Manpower Data Center information supplied in 2009, for years of service data; DoD fiscal year 2009 composite pay and reimbursement rates, for pay factors.

NOTE: Our descriptions of duties and responsibilities are generalized; we recognize that these factors can vary widely with each billet.

[a] Fiscal year 2009 military composite standard pay and reimbursement rates from the Navy.

In general, senior enlisted personnel have served longer than junior officers, and many have attained bachelor's degrees, received advanced technical training, become capable leaders and managers, managed and led comparable numbers of personnel, and been given positional and grade authority over subordinates.

Overall, the general duties and responsibilities of senior enlisted personnel compare favorably with those of junior officers, as they are similar in scope and nature. We conclude that senior enlisted personnel would be best suited for junior officer substitution in billets at grades O-4 and below. However, because the general duties and responsibilities of each billet vary, a billet-by-billet assessment is necessary to determine the efficacy of a substitution.

Based on our assessment of these factors, we conclude that it would be challenging to fill senior officer positions (those designated for O-5s and above) with senior enlisted personnel (or WOs and LDOs). Substitution of senior enlisted personnel in positions designated for O-4s and below is a better fit. Today, WOs and LDOs bridge the gap between the officer and enlisted force.

Criteria to Assess Manpower Alternatives

Existing guidance on manpower mix provides overall direction to the services in determining the workforce mix. To derive criteria for assessing alternative manpower mixes, we reviewed this guidance and discussed its implications with the sponsor. Specifically, we sought to develop a template or process for assessing opportunities for alternative manpower solutions for grades O-4 and below in specific billets.

DoD's existing manpower mix guidance is used to determine whether a DoD position is to be filled by a military service member or by a civilian. The guidance lists specific criteria that dictates incumbency by a military person rather than a DoD civilian. We tailored these criteria to create a framework to determine whether an enlisted, WO, or LDO (or, in some cases, civilian) could serve in the position. For example, instead of determining whether the position should be filled by a military or civilian incumbent, we modified the criteria to consider incumbency by an enlisted person, WO, or LDO (or civilian) rather than an officer. We also developed secondary criteria, discussed in a later section, that need to be considered if an alternative manpower fill of a billet were deemed appropriate.

Alternative Manpower Options to Be Considered

Figure 4.1 illustrates the decision flowchart model that we used for determining the military workforce mix alternatives for new or unfilled billets. The matrix is a series of hurdles that lead to the use of officer, enlisted, or civilian workers.[1] We review each of its questions in this section.

First, for *a new or unfilled position*, is the work *inherently military*? If not, then managers should evaluate the position for civilian performance using DoD's "Inherently Governmental" guidelines. If so, then managers should proceed to consider the military personnel level needed to fill the position.

[1] Not all of these criteria might be applicable to all billets. Moreover, the criteria are equally weighted as we applied them, and failing to meet just one criterion could disqualify substitution. In actual use, subject matter experts might eliminate or selectively weight individual criteria.

Figure 4.1
Decision Flowchart for Determining Military Workforce Mix: Primary Criteria

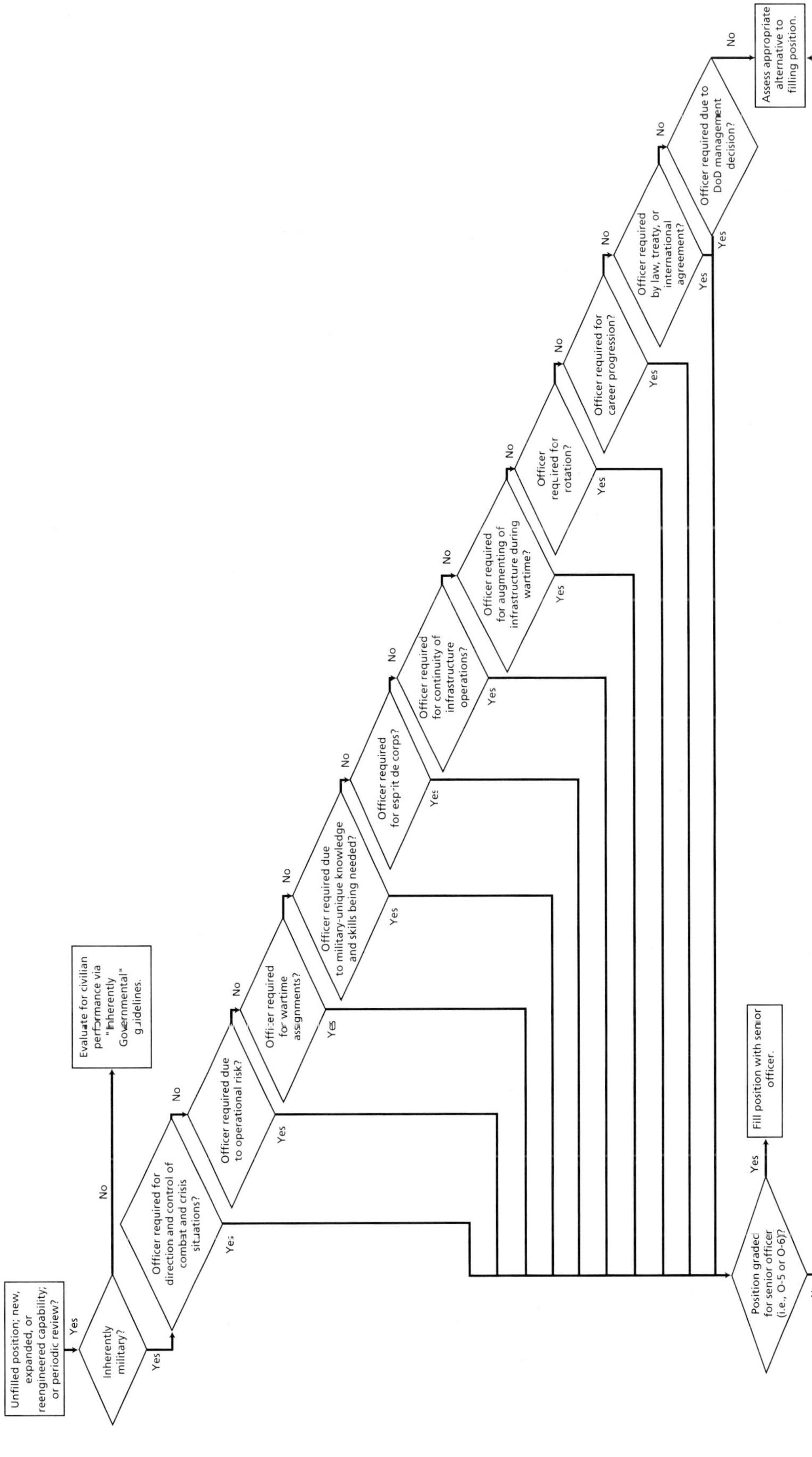

RAND *TR881-4.1*

Second, for inherently military positions, is the position *graded for O-5 or above*? If so, then it should be filled as graded. If not, then managers should consider substituting senior enlisted personnel. We wish to reiterate that we found senior enlisted personnel roughly comparable in terms of education and experience to officers of grades O-4 and below, but not to those of grades O-5 and above. Hence, we suggest that substitution possibilities be considered only for lower grades.

Third, is an officer needed for *direction and control of combat and crisis situations*? Officers are required in certain tactical, operational, and strategic assignments. If the position involves these assignments, then an officer should fill it.

Fourth, is an officer necessary due to *operational risk*? The training an officer receives provides unique skills and abilities not found in other manpower sources. If this condition is present, then the position should be designated for officer performance and not considered for substitution.

Fifth, is an officer needed for *wartime assignments*? A sufficient pool of officers must be available for critical wartime assignments. If the position must be filled by an officer during wartime, then it should be designated for officer performance and not considered for substitution.

Sixth, is an officer needed due to *military-unique KSAs* necessary for the position? If so, then the position should be designated for officer performance and not considered for substitution. An officer's judgment and insight in making informed decisions precludes reliance on other manpower categories.

Seventh, is an officer necessary for *esprit de corps*? If so, then the position should be designated for officer performance and not considered for substitution. Officers are needed to foster public support of recruiting and retention objectives. The need for officers in specific billets for this purpose would vary from position to position.

Eighth, is an officer needed for *continuity of infrastructure operations*? Some officers may occupy billets that require them to perform wartime assignments. The same billets may also be needed to support a DoD capability to train personnel in critical skills. Such positions should remain graded appropriately for officers who are specifically capable of training personnel in critical skills. If this condition is present, then the position should be designated for officer performance and not considered for substitution.

Ninth, is an officer needed to *augment infrastructure during wartime*? During wartime, augmentees often are required for infrastructure support. Many billets are officer-unique billets due to acquired KSAs. If this condition is present, then the position should be designated for officer performance and not considered for substitution.

Tenth, is an officer billet required for *rotation*? Overseas billets requiring officers also require a sufficient base of officers for rotations back to the United States. If this condition is present, then the position should be designated for officer performance and not considered for substitution.

Eleventh, is the position necessary for officer *career progression*? That is, is it a position used to develop unique military competencies? If so, then the position should be designated for officer performance and not considered for substitution.

Twelfth, is an officer required for the position by *law, treaty, or international agreement*? Often, status of forces agreements and North Atlantic Treaty Organization positions require manning by officers. Such positions are not suitable for substitution.

Thirteenth, is an officer needed for the position due to a *DoD management decision*? For some positions, the services may wish to exercise discretional authority in determining whether an officer should occupy the billet. This would preclude substitution.

Negative answers to each of the above questions indicate that a billet could potentially be filled by a senior enlisted rather than an officer. Positive answers to any of the above questions trigger assessment of whether the position is graded correctly. Positions that are correctly graded are not eligible for substitution; incorrectly graded positions may be eligible. Positions that, through this process, are deemed eligible and appropriate for substitution are next considered in light of secondary criteria.

Secondary Criteria

We developed secondary criteria by conducting a literature review of service guidance, by evaluating public and private industry approaches to meeting manpower demands, and through discussions with the sponsor. Although the primary criteria we developed were aligned with DoD manpower mix criteria, the secondary criteria we discuss in this section are those conditions needed to support the utilization of alternative manpower. That is, these secondary criteria address changes that might be needed to support the use of alternative manpower. We used these criteria to evaluate alternative manpower options for the selected specialties that we analyzed.

Once a position is deemed suitable for substitution, managers must consider a set of secondary criteria for determining the range of options for substitutions. These criteria may be framed in terms of the following question:

- Is it suitable to substitute an enlisted for an officer? That is, is it possible to change the workforce type of the person who performs the duties of the position?
- Can the role or responsibility of the billet be changed to meet available manpower? In this case, we consider whether it is possible to modify a portion of the role or responsibilities of the job in order to accommodate the personnel available to fill the position. If there are duties that a person cannot perform, could those duties be shifted to another billet?
- Can WOs or LDOs perform the responsibilities of the position? That is, can a technical expert perform the positional responsibilities?
- Is it possible to do a grade substitution at the billet level? That is, can an individual of lower grade perform the duties of the position?
- Can civilians perform the work? That is, are there capable civilians available to perform the work, and is the work suitable for civilians?

Figure 4.2 shows how these secondary criteria may be applied. Following is a discussion of the secondary criteria.

Figure 4.2 shows that, even if a position meets all the primary criteria for substitution, it might still require an O-4 for two reasons: the position involves *supervision of other officers up to the O-4 level* and requires *collaboration with other field-grade officers*, especially from other nations.

If the position requires neither supervision of nor collaboration with other field-grade officers, then managers should consider a further set of screens that identify whether *other services or public or commercial industry uses a lower-cost labor source for the position*.

Figure 4.2
Decision Flowchart for Determining Military Workforce Alternatives: Secondary Criteria

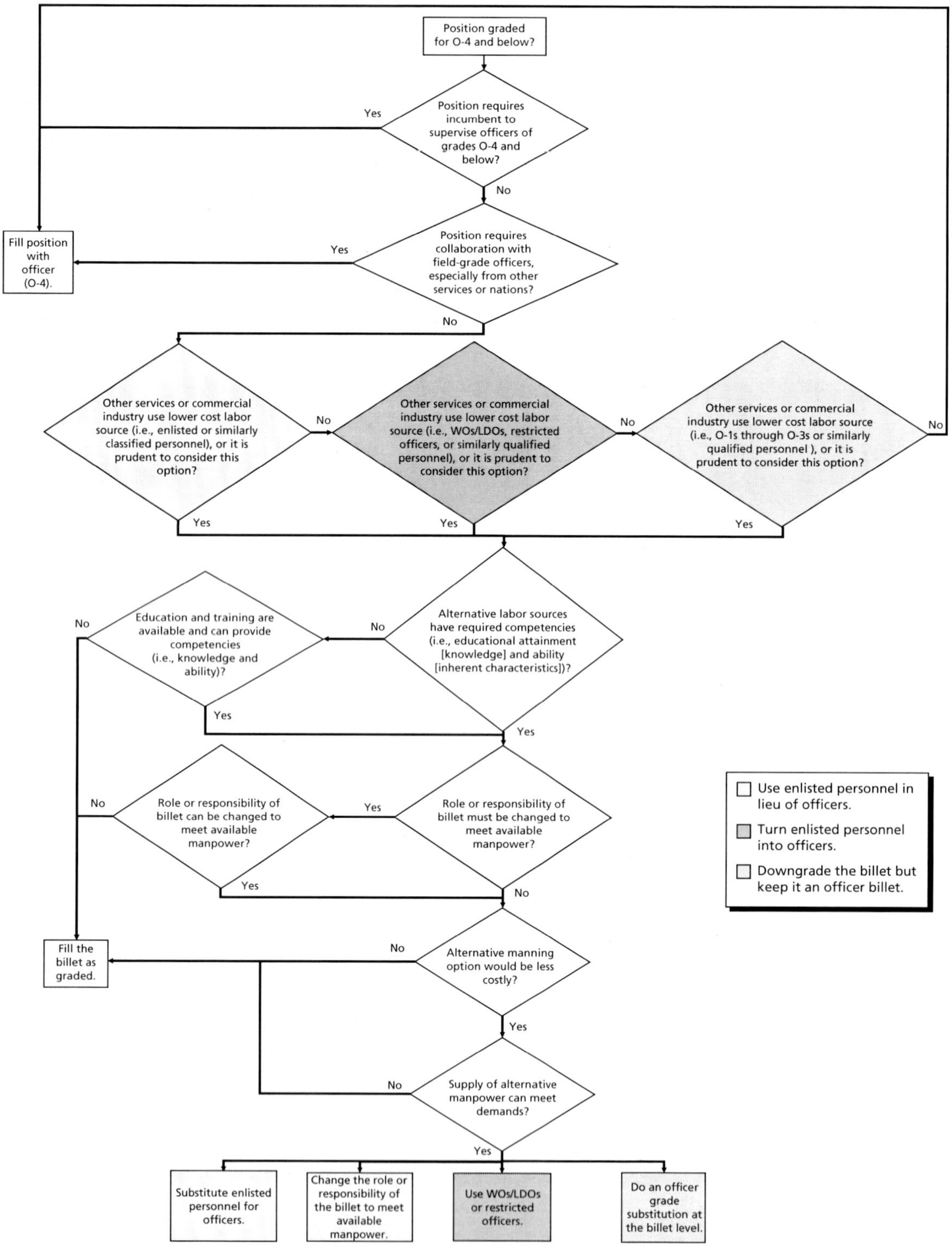

RAND *TR881-4.2*

To compare military occupations with civilian and other government occupations, we reviewed Department of Labor and Office of Personnel Management websites, looking at private and public sector education and training data.[2] This helped us determine the requisite KSAs and experience associated with other occupations. We gathered information on comparable private and public sector jobs, including occupation descriptions and vocational training requirements. We also found data on median income, the size of the field, and expected growth.

We discussed appropriate criteria to consider with the sponsor. We also provided comparisons of public and private training and education needs for comparable occupations, and the sponsor agreed that this was reasonable approach to take and that these factors could be included as criteria.

If the position is one that other military services or private and public sector employers fill with a different labor source, managers should consider whether the alternative labor source has the *prerequisites (e.g., educational attainment) and skills* required to perform the military position. This might include consideration of whether other military services, other government agencies, and private industry use enlisted personnel, WOs, LDOs, junior officers, or other comparable personnel to perform the work. If so, then the position could potentially be filled with an alternative source of manpower. If not, then position should be designated for O-4 performance.

If these other sources of manpower do not have the needed credentials, knowledge, or experience, then policymakers should consider whether additional *education and training could provide these required features.* If not, then the position should be filled as graded.

For positions in which alternative labor sources have or can obtain necessary competencies and skills, policymakers should consider whether *the role or responsibility of the billet must or can be changed to meet available manpower.* If it cannot be changed, then it should be filled as graded.

For positions in which the role or responsibility of the billet can be changed to meet available manpower, policymakers should consider whether the *alternative manning option would be less costly.* If not, it should be filled as graded. If so, then the final consideration for policymakers is whether the *supply of alternative manpower can meet demand.* If not, the position should be filled as graded; if so, then managers may choose a substitution option to fill it.

---

2 Department of Labor, Bureau of Labor Statistics (2009); Office of Personnel Management (undated).

CHAPTER FIVE

Application of Criteria to Selected Occupations

Once we identified primary and secondary criteria for determining whether a position is suitable for substitution, we sought to apply our criteria to select positions to show how the process works. This exercise demonstrates factors that DoD may wish to consider in assessing the possibility of alternative manpower mixes for vacant officer positions. It is not an exhaustive list of such considerations, and our select positions are not an exhaustive listing of positions for which DoD might wish to pursue alternative manpower mixes.

In this chapter, we first described how we identified officer vacancies for further consideration. We then present a broad overview of our substitution suitability analysis for each position. Finally, we summarize the findings of our position-by-position analysis and their implications for future work.

Officer Vacancies as Identified by the Services and the Combatant Commands

The project sponsor provided us with an initial set of data on vacancies that had been collected via a data call to the services and the COCOMs. Each service listed its critical skill shortages, both enlisted and officer, along with inventory and requirement counts. The COCOMs generated a document that listed, by service, grade, number of manpower authorizations and assigned personnel, and DoD occupation code, critical manpower shortages. Using these data, we calculated the gap between authorized and assigned officer personnel and the fill rate for each occupation.

To select occupations for further analysis, we combined the service and COCOM data into a master spreadsheet. In conjunction with the sponsor, we used three criteria to select occupations for further consideration. First, we arrayed occupations by the number of unfilled officer billets (i.e., the absolute size of the gap between required and available officers, by occupation). The occupations that were selected for analysis were those with current shortages. Second, using data from the Defense Manpower Data Center's Forces, Readiness and Manpower Information System, we calculated the ratio of enlisted personnel to officers for all given vacant officer occupations and compared these ratios with the average ratio for each service. Higher ratios might indicate positions in which officers are performing duties that enlisted personnel, who are available in greater numbers, might be able to perform instead. Of course, we also considered positions that could potentially be filled by WOs, LDOs, and civilians. Third, we identified occupations that both the services and the COCOMs identified as having critical skill shortages. We considered similar military occupational specialties (MOSs) and specialty codes when assessing appropriate substitution for a vacant officer MOS/Air Force Specialty Code.

Analysis and discussion with the sponsor led us to select the following occupations for further analysis:

- **UAS operators.** The sponsor suggested this occupation for review. As previously noted, the services use different types of manpower to pilot UASs for "similar" missions. Although they employ different types of manpower, the services are acquiring similar UAS airframes.
- **Air Force communications and information officers.** We chose to analyze this occupation because both the services and the COCOMs indicated that there is a challenge in manning these officer billets. Some services and COCOMs also indicated that there is a challenge in manning enlisted billets in this occupation.
- **meteorological and oceanographic (METOC) officers (weathermen).** The sponsor suggested this occupation for review. There is only a small number of METOC officers in the services and the COCOMs, and one service identified METOC officers as a manning challenge.
- **Army telecommunications officers.** Both the Army and the relevant COCOMs identified this position as a manning challenge for officers and, in some instances, even for enlisted personnel.
- **Army information systems officers.** Both the Army and the relevant COCOMs identified this position as a manning challenge for officers and, in some instances, even for enlisted personnel.
- **Army signal officers.** Both the Army and the relevant COCOMs identified this position as a manning challenge for officers and, in some instances, even for enlisted personnel.

In some cases, we found that the services required an officer in positions that were performed by civilian workers with skills similar to those of enlisted personnel. In other cases, we found that there was a good match between civilian and military job requirements and the type of worker filling the position. The next section presents an overview of our analysis of the potential for substitution in each of these positions.

Review of Positions

We reviewed positions in each of these six occupations against the primary and secondary criteria presented in the previous chapter. To perform this assessment, we reviewed public and private manpower approaches; conducted a literature review of the selected occupations; interviewed subject matter experts; and added insights based on our own knowledge, experience, and judgment. Tables 5.1 and 5.2, respectively, outline the analysis for the primary and secondary criteria.

UAS Operators

Whether the UAS operator position is inherently military depends on how the system is operated. Some UAS unit organizations, such as those in the Navy, include an officer assigned overall responsibility and who serves as the mission commander for direction and control. Under

Table 5.1
Primary Criteria Applied to Selected Occupations: Summary of Results

Primary Criteria	UAS Operators	Air Force Communications and Information Officers	METOC Officers	Army Telecommunications Officers	Army Information Systems Officers	Army Signal Officers
Is the position inherently military?	Debatable; the Central Intelligence Agency uses civilians to pilot UAS aircraft	Some positions converted to civilian billets	Debatable; senior officers want senior METOC officers for military operations	Yes for O-3 billets, no for test/evaluation O-4 billets	Yes for most billets	Yes
Is an officer required for direction/control?	Mission commander only	At COCOM, an officer is required	No	No	No	No
Is an officer required due to operational risk?	Service discretion applies	Some billets may have no risk	Service discretion; options are available	No	No	No
Is an officer required for wartime assignments?	Situation dependent	Officers needed, but not in all cases	No	No	No for O-3 billets, yes for O-4 billets	No
Is an officer required due to military-unique KSAs?	No	Yes in some domains, no in others	No	No	No	No
Is an officer required for esprit de corps?	No	No	No	No	No	No
Is an officer required for continuity of operations?	Depends on the mission	No	No	No	No	No
Is an officer required to augment infrastructure?	Depends on the mission	Yes in some cases, no in others	Depends on the mission	No	No	No
Is an officer required for rotational purposes?	Depends on the specialty structure	Depends on the number of billets in CONUS vs. outside CONUS	Depends on the specialty structure	No	No	No
Is an officer required due to career progression?	Depends on the specialty structure	Depends on retention and structure	Depends on the specialty structure	Yes	No	No
Is an officer required due to law, treaty, or international agreement?	Requires a pilot in some operational settings	No	No	No for O-3 billets, maybe for O-4 billets	No for most billets	No
Is an officer required due to DoD management decision?	Service discretion applies	Service discretion applies	Service discretion applies	No for O-3 billets, yes for some O-4 billets	No	No

NOTE: CONUS = continental United States.

Table 5.2
Secondary Criteria Applied to Selected Occupations: Summary of Results

Secondary Criteria	UAS Operators	Air Force Communications and Information Officers	METOC Officers	Army Telecommunications Officers	Army Information Systems Officers	Army Signal Officers
Supervises O-4s and below?	Depends on unit organization	Likely just junior officers and civilians	Few billets supervise officers	No	No for O-3 billets; yes or maybe for O-4 billets	No for most, yes at regiment level
Collaborates with field grades?	Yes, with other services; common training	Yes for COCOMs, no for in-house billets	Yes for COCOMs; reach-back capability	Yes, but manageable	Some, but manageable	No for some billets, yes for others
Others use lower grades, WOs, or LDOs?	Yes	No, but others use civilians	Yes, Navy uses LDOs	Yes	Yes, LDOs/WOs	Yes
Do alternative sources have KSAs?	Yes	Civilians have KSAs	Depends on work; O-4s and higher, those with advanced degrees	Yes	Yes for O-3 billets, maybe for O-4 billets	No
Are education and training available?	Yes, at the UAV training center	Yes in civilian sector	Navy aerographer's mates do METOC work and are trained	Probably (education); yes (training)	Yes	Yes
Must the job role change?	No	No	Needs review	No	Maybe for O-3 billets, no for O-4 billets	No
Can the job role be changed?	N/A	N/A	Needs review	Maybe or no	Yes	Yes, as needed

Table 5.2—Continued

Secondary Criteria	UAS Operators	Air Force Communications and Information Officers	METOC Officers	Army Telecommunications Officers	Army Information Systems Officers	Army Signal Officers
Will alternative manpower be less costly?	Yes	Civilians may be cheaper	Yes	Yes	Yes	Yes
Can the supply of alternative manpower meet demand?	Long-term supply must be reviewed	Long-term supply must be reviewed	Long-term supply must be reviewed	Long-term supply must be reviewed	WOs may be able to fill O-3 billets; junior officers may be able to fill O-4 billets	Yes for some, unknown for others
Is it possible to substitute an enlisted for an officer?	Currently done in the Army and the Navy		Yes, or use civilians	Use senior NCOs as network/systems engineers	Yes	Use NCOs for some billets
Is it possible to change the role of billet?	An officer is a mission commander; an enlisted is an operator	Use civilians	Potentially	N/A	Potentially	N/A
Is it possible to use WOs or LDOs ?	N/A	N/A	Yes	Yes	Yes	Use WOs for some billets
Is it possible to do a grade substitution?	Air Force has shifted to nonrated	Some may be feasible	N/A	Substitute O-3s for O-4s	Yes for some	Yes, at brigade, division artillery, and regiment S-6 levels

NOTE: UAV = unmanned aerial vehicle.

the mission commander are an enlisted UAS pilot who would fly the aircraft and a sensor operator who would operate onboard cameras and other airframe equipment. Although rated pilots are required in some settings, they may not be required in all circumstances. The UAS pilot does not supervise officers, and the Army offers extensive training for enlisted personnel who fill that position. Possible manpower alternatives for UAS pilots include substituting enlisted personnel for officers, changing the role of the billet (e.g., using an officer as the mission commander but an enlisted as the UAS operator), using nonrated officers, and even using civilians in remote-split operations (i.e., having civilians within the United States operate a UAS in an overseas theater).

Air Force Communications and Information Officers

Civilians perform this job outside the military, but not necessarily in a military context. We found that some Air Force units have converted this specialty to a civilian billet. Some of these officer positions require a military incumbent, and a specialty force structure is required for rotation. COCOM billets, but not service billets, require officers to collaborate with officers from other services. Civilians do have the capability to meet the demands of this position, and civilians without the necessary domain knowledge might be able to obtain it through experience or training. Therefore, we suggest that civilians be considered in manning alternatives.

METOC Officers (Weathermen)

It is enlisted personnel (particularly Navy aerographer's mates) rather than officers who do most weather analysis in the Navy. Perhaps the two key issues to consider in alternative workforce mixes for this position are how weather may affect the operational mission and who would be in the best position to advise the COCOM. Technological improvements, satellite imagery, and increasing capabilities in weather forecasting provide great reach-back capability, and large weather centers already prepare and disseminate weather predictions. Senior METOC officers and their civilian counterparts have advanced degrees. Nevertheless, the Navy also uses LDOs in this position. Our analysis suggests that alternative manning sources for METOC officers might include the increased use of LDOs, the substitution of some lower-grade officers for those of higher grades, the use of enlisted personnel in some positions now held by officers, and the increased use of civilians.

Army Telecommunications Officers

Army telecommunications officers are technical experts who "engineer, design, develop, install, implement, integrate, test, accept, and upgrade tactical, strategic, and sustaining base wired and wireless telecommunications systems and networks enterprise-wide."[1] The Army currently limits entry into this career field to branch-qualified signal officers with 4–7 years of service.

There are about 220 officers of all grades in this functional area. Senior NCOs might be able to substitute for junior officers in about 60 positions, while junior officers could substitute for more-senior officers in perhaps seven positions.[2] For all these positions, the junior personnel would require additional education and training.

---

[1] Army Pamphlet 600-3.

[2] Such positions include O-3 network engineers working in division G-6 active component sections, of which there are no more than ten, and O-3 systems engineers working in signal battalions or brigades.

Senior NCOs working under the supervision of a capable senior officer might be able to substitute experience and military education and training for postsecondary education in some billets. Senior signals NCOs (E-7s and above) are currently responsible for several network planning and engineering tasks, indicating their capacity in this area.

Substituting NCOs for junior officers would, however, reduce developmental opportunities for O-3s, limiting the future supply of O-4s. However, the substitution is cost neutral. NCOs with the requisite experience and training have an economic life cycle cost to the federal government comparable to that of junior officers. An E-7 with 13 years of experience (the average level of experience at promotion to that grade) has an economic life cycle cost to the federal government comparable to the cost of an O-3 with six years of experience.[3]

It might be more feasible to regrade some O-4 systems engineers in signal brigades to O-3 or WO positions. The number of experienced signal officers concentrated in Army telecommunications units could compensate for any defect in experience in tactical or operational matters. There are approximately seven such positions.

Army Information Systems Officers

These officers "integrate diverse forms of enterprise services such as configuration management, active directory, database management, email, and Web portals that must operate seamlessly in support of the Army and Joint, Interagency, Intergovernmental, and Multinational . . . operations worldwide."[4] There are approximately 600 officers in all grades in this specialty. Officers in this functional area must collaborate with a wide variety of information-system users to plan and integrate the use of various software products. Positions in which the incumbent must collaborate with field-grade officers to develop policy or operational planning are not amenable to substitution.

Positions involving the technical management and maintenance of automation systems might be amenable to some degree of substitution. Because of the high degree of technical competence required, only WOs should be considered as substitutes for junior officers in these billets. WOs often have responsibilities similar to those of junior commissioned officers.[5] There are about 110 Army technical management and maintenance of automation system positions that might be amenable to substitution. These are distributed across six types of billet: information systems management officer, brigade combat team S-6 section (not the staff primary), signal brigade or command, division-level G-6 section, staff positions in the Network Enterprise Technology Command/9th Signal Command (Army), and positions on the Army or Joint Staff. Junior officers might also function effectively as the S-6 for support brigades.

---

[3] Dahlman (2007) argues that neither the annual Regular Military Compensation cost nor the DoD Comptroller's Standard Military Composite Rates are the proper figures to use in analyzing billet conversions. His method incorporates the annual Regular Military Compensation cost for an enlisted or officer at each year of service plus a probability-adjusted value for future retirement and Tricare for Life benefits for that year of service. Because officers at any early year of service have a much higher probability of serving to retirement than do enlisted personnel at the same early year of service, their comparable costs at an equal year of service will always be higher, partially because of higher Regular Military Compensation costs but primarily because of the probability-adjusted future retirement and Tricare for Life benefits. However, because actual billet conversions typically substitute a more experienced enlisted person for a more junior officer and involve personnel with unequal levels of service, the costs can in fact be similar, and the substitution can thus result in no cost savings.

[4] Army Pamphlet 600-3, p. 223.

[5] Army Pamphlet 600-3, p. 204.

As with telecommunications systems engineers, such substitution should be limited to units in which other signal officers could provide expertise in tactical and operational considerations. Substituting WOs for junior officers would probably be more cost-effective than using NCOs.

Army Signal Officers

Army signal officers are responsible for planning and integrating communications systems support into Army and joint operations. There are more than 2,500 such officers in all grades. Such planning and integration frequently requires collaboration with field-grade officers from other specialties, and an understanding of tactics or operations is therefore often more important than technical competence within the domain. In such cases, substituting junior officers, WOs, or NCOs for more-senior officers is not warranted.

Positions for which substituting NCOs for junior officers might be appropriate include battalion S-6, service school instructor, and ROTC assistant professors of military science. Battalion signal officers usually manage a fairly narrow range of capabilities and typically operate in standard patterns. It would be advisable to limit such substitution to units with relatively simple missions (e.g., support brigades). Service school instructors are often NCOs, regardless of the students' grade; such instruction, however, must focus on systems' technical capabilities. Finally, signal NCOs could provide human resources and administrative support as assistant professors of military science to enable ROTC detachments to function. There is a maximum of about 210 such positions that might be amenable to substitution.

There are three types of O-4 billets for which junior officers, WOs, or NCOs might be substituted. Senior NCOs could serve effectively as doctrine developers or training developers; several retired NCOs are currently serving in this capacity at the Signal Center. WOs or junior officers might also be able to substitute for O-4s as the S-6 for support brigades and similar units. There is a maximum of 47 such billets that our analysis suggests could be converted.

Future Considerations

We conducted a broad assessment of positions in the selected occupations. A more detailed billet-by-billet examination would be necessary to evaluate the appropriateness of alternative manpower options. Subject matter experts would best perform this examination. For example, a service community or career-field manager's understanding of incumbent responsibilities would be necessary to judge the demands of the position and the grade, specialty, and experience required. In addition, the use of alternative manpower might have cascading or unintended effects that we have not assessed. For example, the increased use of enlisted personnel might cause some enlisted positions to be understaffed.[6] The evaluation of alternative manpower sources should be an iterative process performed frequently to capture changes in the work and workforce.

Although some billets in these selected occupations seem to be promising candidates for alternative manning, other billets may not be. Service manpower experts would need to con-

---

[6] The services are already taking advantage of higher levels of education and experience within the force, and substituting senior enlisted personnel for junior officers could have broader implications for the enlisted force in terms of accession, retention, and force management profiles. See, for example, Hansen and Wenger (2003).

duct a detailed, billet-by-billet examination of billets designated for officers levels O-1 through O-4. However, our assessment shows that substitution of alternative manpower sources is feasible in theory.

Billets are unique, and their training, experience, and work requirements often vary by service and position and even over time. Some billet characteristics, including the grade, experience, leadership, and collaboration required, are more important than others in considering substitution possibilities. Some conditions may make billet conversion easier, more difficult, or impractical. Table 5.3 lists factors that can make conversion easier or more difficult. These factors should apply to all the services equally, although, once again, their importance may vary.

Table 5.3
General Factors Affecting Conversion

Factor	Conversion Is Easier	Conversion Is More Difficult
Grade	Senior enlisted (E-7 through E-9) substituting for O-4 positions and below	Enlisted substituting for O-5 or O-6 positions
Cost	Less expensive	More expensive
Education level	Education and knowledge acquired without a bachelor's degree	Bachelor's degree or higher required
Years of service	Similar experience requirements	Different, more-extensive, and discrete experience requirements
Training	Certification or education requirements that are prerequisites for a position can be achieved by enlisted personnel through a readily available training track; KSAs can be reasonably met without vast changes; training is achievable by enlisted personnel; same standards	Bona fide certification or education requirements are prerequisite for position; KSAs cannot be achieved, or it is infeasible or more costly for enlisted personnel to achieve these KSAs on the training track
Nature of work	Work that can be demonstrated with a measurable output; differences between officer and enlisted performance are minimal	Work requiring experience and decisionmaking based on deep KSAs acquired through officer positions
Scope of work	Staff, administrative, policy, or operational/tactical with appropriate oversight/decisionmaker available as required	Strategic
Level of responsibility	Staff or operational position that produces output; can have policy implications	Position with specific policy responsibilities commensurate with grade
Span of control	Small span of control; immediate work decisions	Large span of control
Unit size or organization	Large organization, where an enlisted can confer and consult with officers, or a small organization, where there is a big impact if the billet is unfilled (e.g., if there are only two billets for a skill and one is unfilled, then the impact of an unfilled billet is greater)	Large or small organization; independent work; complex unit(s)
Work value	Operational output	Training output
Command/NJP authority	Not required for the position	Required for the position
Subordinates supervised	Commensurate with grade	Inappropriate for grade

Table 5.3—Continued

Factor	Conversion Is Easier	Conversion Is More Difficult
Workforce interactions	Horizontal coordination	Vertical coordination
Workforce identity	Weak or changeable identity and culture	Strong identity and culture
Position type	General duty, measurable output	Organizational position
Development of incumbent	The position not designated as one needed for furthering the development of junior officers	The position is used to provide officers with experience needed for higher positions
Enlisted force structure	The career field is bottom heavy; there is a surplus of lower-grade billets to grow personnel	The career field is top heavy; it is hard to grow additional NCOs/senior NCOs without larger growth in lower grades
Officer force structure	The officer career field is top heavy; this may be the cause of the challenge; the conversion of billets to enlisted would help	The career field is bottom heavy; there should not be a shortage as company-grade officers move to field grade
Qualification	Entry requirements are met by enlisted applicants	Entry requirements are difficult to meet

Training that is readily available and can be reasonably achieved by enlisted personnel would support conversion efforts. Discrete training requirements or those that cannot be easily met would make conversion more difficult.

Measurable work would support conversion. Work requiring experience and decision-making based on deep KSAs acquired through officer positions would be more difficult to convert.

A scope of work that is staff, administrative, executing policy, operational or tactical in nature and that has appropriate oversight or makes a decisionmaker available would support conversion. If the scope of work can be executed by staff members with oversight, or is administrative in nature, then it would support conversion. If the scope is strategic, then an officer would best fill the position.

A level of responsibility reflecting that of a staff or operational position providing an output or having policy implications would support conversion. A position with responsibilities for *establishing* policy may not be appropriate for conversion, but one whose performance might have implications for *executing* policy might be. A position requiring specific policy or other responsibilities commensurate with the established grade would make conversion more difficult.

A large span of control would make it difficult to convert a position. A small span of control that is needed only for immediate work decisions could support conversion.

The effect of organization size on conversion potential varies. Senior enlisted personnel could perhaps more easily fill vacant billets in large organizations, especially if they are able to confer and consult with officers in similar positions. At the same time, small organizations with a vacant position would benefit from having a senior enlisted person fill the billet because the adverse impact of leaving the billet unfilled would be great. No matter the size of the organization, if the work is independent and complex, then the position may be difficult to convert.

If a position produces an operational output, then it may be easier to convert than one with a training output. This is in part because a training output may be more closely linked to an officer's career field and experience obtained in the career track.

If the position is one in which command and NJP authority are needed, then the billet is best filled by an officer.

If subordinates are commensurate with the incumbent's grade, then the position would support conversion. Having higher-grade subordinates would make conversion more difficult.

Similarly, horizontal coordination (i.e., coordination between an incumbent and subordinates that is more collaborative) can support conversion. Vertical coordination (i.e., work an incumbent is more likely to direct rather than collaborate on) may make conversion more difficult.

A position with a weak or changeable identity and culture would be easier to convert than one with a strong identity and culture. Those working with a position with a strong identity and culture may be accustomed to seeing an officer in the billet.

A general-duty billet that has a measurable output would be easier to convert from officer to senior enlisted than one that is an organizational or command position.

The role of the position in developing junior officers can also affect conversion possibilities. Billets that are designated as an experience tour and necessary for developing junior officers would be more difficult to convert. Those not considered necessary or that provide minimal value in providing a junior officer with experience or development would be easier to convert.

Enlisted and officer force structure affect conversion in conflicting ways. A bottom-heavy enlisted structure (i.e., one with a surplus of lower-grade billets) would support conversion, while a top-heavy enlisted structure would make it more difficult to develop more-senior enlisted personnel to support converted positions without still more growth in the lower grades. Conversely, a top-heavy officer structure may be the cause of the challenge in filling billets—a challenge that using senior enlisted personnel could alleviate.

Finally, the broad qualification requirements for a position can support or hinder its potential for conversion. A position with entry requirements that are easier to meet satisfactorily will, of course, be easier to convert than one with qualifications that are more difficult to meet.

Overall, our assessment indicates that it would be easier for senior enlisted personnel (E-7s though E-9s) to substitute at O-4 positions and below than to substitute in billets at more-senior officer grades. If a converted billet were deemed to be less costly, then it would be a more attractive option to pursue. If the education and knowledge needed for a position could be obtained through experience, then conversion for that position would be easier than that for a position requiring a bachelor's degree or higher. If a billet requires similar levels of experience or number of years of service, then it would be easier to convert than a billet whose experience requirements were vastly different or discrete.

CHAPTER SIX

Conclusions and Recommendations

In this work, we explored opportunities to enhance the junior officer (O-4 and below) personnel fill rates of the services and COCOM headquarters through the use of alternative mixes of military manpower for particular billets. The possibility of doing this is one of the benefits of having an enlisted force that is more senior, more experienced, more educated, and more capable than at any other time in the history of the U.S. military. It is also one of the benefits of today's highly capable WO, LDO, and civilian workforce.

The ever-increasing capabilities of the WO, LDO, civilian, and enlisted workforces have supported an increase in such personnel's duties and responsibilities, including performing in positions previously held only by line officers. USC, OSD, and CJCS guidance provides little direction or criteria to the services in regard to designating a billet as officer or enlisted, and the services have begun to expand the responsibilities of their capable workforce. The services govern the workforce designation of specific billets, and their culture, policy, and practices play a role in determining the workforce type selected.

Service processes for evaluating and classifying jobs determine the occupation, type, and grade appropriate for performing billet tasks, and mission demands typically drive the requirements of a billet. However, as we have shown, the services sometimes take different approaches filling billets, even billets involving similar missions and equipment.

To be sure, there are positions that only officers can fill; likewise, there are positions that are best filled by enlisted personnel. Nevertheless, there are an increasing number of ambiguous areas. Senior enlisted personnel, WOs, and LDOs (and civilians) have increased experience and education, and the services are beginning to increase the responsibilities of these personnel, asking them to perform in billets currently classified for performance by line officers. This research provides a process to help leadership determine whether such opportunities for substitution exist in other positions.

The services consider many variables in filling billets. For example, they reserve some billets based on the culture of the position or to ensure a sufficient base of junior officers for higher-graded positions. The services may be wary of substituting enlisted personnel in positions used to develop officers.

The manning process for determining military positions is dynamic and merits continuous review. Options do exist, and they should be considered in filling vacant officer positions with alternative manning sources. Databases are available to support comparisons between military work and similar work in both the private sector and elsewhere in the public sector. In fact, DoD has created crosswalks between military occupations and the Department of Labor occupational system. DoD can tailor its manpower mix criteria to address alternative options for meeting positional demands.

There has been some conversion of junior-graded positions, but mostly from officer to WO or LDO positions. WOs and LDOs can meet the technical, education, and experience requirements of some positions and can bridge the gap between officer and enlisted positions. Bachelor's degrees are desired but not required for WOs and LDOs in these positions. Senior enlisted educational attainment has increased substantially in the last 20 years. This suggests that there may be further opportunities for substituting enlisted personnel for officers (especially for substituting senior enlisted personnel for junior officers) and for conducting further substitution of WOs and LDOs for officers. In addition, there are opportunities to use civilians in officer positions.

We recommend that DoD further seek to leverage the experience, education, and capability of the enlisted, WO, LDO, and civilian workforces. OSD should promulgate clear guidance for designating positions for performance by enlisted personnel, WOs, LDOs, or officers. DoD should also revise DoD Instruction 1100.22, *Guidance for Determining Workforce Mix*, to provide criteria and a process for the services to determine whether the military person is to be officer, enlisted, WO, or LDO. OSD should direct the services to evaluate new or vacant funded positions and periodically reevaluate existing positions using revised guidance.

APPENDIX

Guidance on Officer/Enlisted Manpower Categorizations

This appendix is organized into seven sections: USC, OSD, CJCS, Army, Navy, Air Force, and Marine Corps. In each section, we review both general guidance and guidance pertaining to the military, officer, enlisted, and civilian workforce types. Guidance, either directly quoted or paraphrased, is in roman type; commentary or analysis is in italic type.

United States Code

General Guidance

10 USC 129a. General personnel policy. The Secretary of Defense shall use the least costly form of personnel consistent with military requirements and other needs of the Department.

10 USC 805. Except as otherwise specifically prescribed by law, the Air Staff shall be organized in such manner, and its members shall perform such duties and have such titles, as the Secretary may prescribe.

10 USC 807, Sec. 8067. Other functions in the Air Force requiring special training or experience shall be performed by members of the Air Force who are qualified under regulations prescribed by the Secretary, and who are designated as being in named categories.

10 USC 10102. [The mission of the Reserve component is to] provide trained units and qualified persons available for active duty in the armed forces, in time of war or national emergency and at such other times as the national security may require.

Military

10 USC 165. Responsibility of Secretaries of Military Departments.--Subject to the authority, direction, and control of the Secretary of Defense and subject to the authority of commanders of the combatant commands under section 164(c) of this title, the Secretary of a military department is responsible for the administration and support of forces assigned by him to a combatant command.

The Defense Officer Personnel Management Act (December 1980) and 10 USC 831. *These limit the number of field grade officers and chief and senior master sergeants based on budgeted officer and enlisted end strengths.*

10 USC 12011 and 12012. *These limit grades for Active Guard and Reserve field-grade officer and senior enlisted authorizations. These authorizations are in addition to the overall active Air Force grade ceilings and are not chargeable to the active grade ceilings. Grades for Active Guard and Reserve tour enlisted authorizations are not subject to the limits on overall numbers imposed* *by 10 USC 517 for chief and senior master sergeants; however, they are limited by 10 USC 12012. The DoD authorization bill establishes annual grade ceilings for chief and senior master sergeants.*

10 USC 10211. Within numbers and grades as the secretary of the department may prescribe, Guard and Reserve officers may be on extended active duty (other than for training) as advisors at the seat of government and at headquarters responsible for reserve affairs, to participate in preparing and administering the policies and regulations affecting those reserve components.

10 USC 10305(h). *There are to be no fewer than five officers from both Air National Guard and the U.S. Air Force Reserve serving on the Air Staff with the Air Reserve Forces Policy Committee. In all other ways, the duties and responsibilities of a 10 USC 8021 officer are the same as a 10 USC 10211 officer. These authorizations are usually colonels.*

Foreign Military Sales (FMS) Manpower. *The Arms Export Control Act, 22 USC 2751, and all subsequent sections that relate to that section and the Foreign Assistance Act of 1961, as amended, authorize the sale of defense articles and services. The DoD appropriation and authorization acts include in the U.S. Air Force statutory end strength ceiling manpower resources to support FMS. Although most FMS manpower is reimbursable, the Air Force programs the requirements to ensure coverage in the Future Years Defense Program. Headquarters, U.S. Air Force does not retain a "pool" of manpower authorizations available for new FMS programs. FMS manpower authorizations are full time or part time. Positions performing less than 10 percent FMS work are not considered FMS.*

10 USC 12402. *Air National Guard officers below brigadier general on duty in the National Guard Bureau take part in developing, administering, and coordinating all programs, policies, principles, concepts, and plans for the National Guard. These officers also assist states in organizing and operating National Guard units. The number of Air National Guard officers below the grade of brigadier general cannot exceed 40 percent of the number of Air Force officers in that grade authorized for duty in that bureau. A tolerance limit of one authorization per grade will be allowed to compensate for the small number involved as long as the overall 60:40 ratio is not exceeded.*

Officer

10 USC 501. The term "officer" means a member of the naval service serving in a commissioned or warrant officer grade. It includes, unless otherwise specified, a member who holds a permanent enlisted grade and a temporary appointment in a commissioned or warrant officer grade. The term "commissioned officer" means a member of the naval service serving in a grade above warrant officer, W-1. It includes, unless otherwise specified, a member who holds a permanent enlisted grade or the permanent grade of warrant officer, W-1, and a temporary appointment in a grade above warrant officer, W-1. The term "warrant officer" means a member of the naval service serving in a warrant officer grade. It includes, unless otherwise specified, a member who holds a permanent enlisted grade and a temporary appointment in a warrant officer grade.

10 USC 343, Sec. 3548. Army Duties: warrant officers; limitations. Under regulations prescribed by the President, a warrant officer may be assigned to perform duties that necessarily include those normally performed by a commissioned officer.

10 USC 343, Sec. 166. Combatant Command Staff.

(1) Each unified and specified combatant command shall have a staff to assist the commander of the command in carrying out his responsibilities. Positions of responsibility on the

combatant command staff shall be filled by officers from each of the armed forces having significant forces assigned to the command.

(2) An officer may be assigned to a position on the staff of a combatant command or, in the case of such a position that is designated under section 601 of this title as a position of importance and responsibility, may be recommended to the President for assignment to that position, only –

(A) with the concurrence of the commander of such command; and

(B) in accordance with procedures established by the Secretary of Defense.

(3) The Secretary of Defense may waive the requirement under paragraph (2) for the concurrence of the commander of a combatant command with regard to the assignment (or recommendation for assignment) of a particular officer to serve on the staff of the combatant command if the Secretary of Defense determines that such action is in the national interest.

Authority to Suspend Subordinates. - In accordance with procedures established by the Secretary of Defense, the commander of a combatant command may suspend from duty and recommend the reassignment of any officer assigned to such combatant command.

10 USC 343, Sec. 716. Notwithstanding any other provision of law, the President, within authorized strengths and with the consent of the officer involved, may transfer any commissioned officer of a uniformed service from his uniformed service to, and appoint him in, another uniformed service. The Secretary of Defense, the Secretary of Homeland Security, the Secretary of Commerce, and the Secretary of Health and Human Services shall jointly establish, by regulations approved by the President, policies and procedures for such transfers and appointments.

10 USC 164. Each unified and specified combatant command shall have a staff to assist the commander of the command in carrying out his responsibilities. Positions of responsibility on the combatant command staff shall be filled by officers from each of the armed forces having significant forces assigned to the command.

(2) An officer may be assigned to a position on the staff of a combatant command or, in the case of such a position that is designated under section 601 of this title as a position of importance and responsibility, may be recommended to the President for assignment to that position, only--

(A) with the concurrence of the commander of such command; and

(B) in accordance with procedures established by the Secretary of Defense.

10 USC 164, Sec. 532. Qualifications for original appointment as a commissioned officer

(a) Under regulations prescribed by the Secretary of Defense, an original appointment as a commissioned officer (other than as a commissioned warrant officer) in the Regular Army, Regular Navy, Regular Air Force, or Regular Marine Corps may be given only to a person who--

(1) is a citizen of the United States;

(2) is able to complete 20 years of active commissioned service before his sixty-second birthday;

(3) is of good moral character;

(4) is physically qualified for active service; and

(5) has such other special qualifications as the Secretary of the military department concerned may prescribe by regulation.

10 USC 164, Sec. 801. Article 1. Definitions

In this chapter: The term "commanding officer" includes only commissioned officers. The term "officer in charge" means a member of the Navy, the Marine Corps, or the Coast Guard designated as such by appropriate authority.

Enlisted

10 USC 501. The term "enlisted member" means a member of the naval service serving in an enlisted grade or rating. It excludes, unless otherwise specified, a member who holds a permanent enlisted grade and a temporary appointment in a commissioned or warrant officer grade.

The term "rating" means the name (such as "boatswain's mate") prescribed for members of an armed force in an occupational field. The term "rate" means the name (such as "chief boatswain's mate") prescribed for members in the same rating or other category who are in the same grade (such as chief petty officer or seaman apprentice).

Civilian

10 USC 501, Sec. 1580. The Secretary of Defense or the Secretary of the military department concerned may designate as an emergency essential employee any employee of the Department of Defense, whether permanent or temporary, the duties of whose position meet all of the following criteria:

(1) It is the duty of the employee to provide immediate and continuing support for combat operations or to support maintenance and repair of combat essential systems of the armed forces.

(2) It is necessary for the employee to perform that duty in a combat zone after the evacuation of nonessential personnel, including any dependents of members of the armed forces, from the zone in connection with a war, a national emergency declared by Congress or the President, or the commencement of combat operations of the armed forces in the zone.

(3) It is impracticable to convert the employee's position to a position authorized to be filled by a member of the armed forces because of a necessity for that duty to be performed without interruption.

Office of the Secretary of Defense

General Guidance

DoDD 1100.4, *Guidance for Manpower Management*. National military objectives shall be accomplished with a minimum of manpower that is organized and employed to provide maximum effectiveness and combat power. Manpower requirements are driven by workload and shall be established at the minimum levels necessary to accomplish mission and performance objectives.

DoDI 1100.22, *Guidance for Determining Workforce Mix*. Use the least costly form of personnel (DoD civilian, military, or private contract) consistent with military requirements and other needs of the Department. Workforce mix shall be established to accomplish Defense missions commensurate with acceptable risk. When designating management positions as military (Active or Reserve) or civilian, career opportunities for both categories of personnel shall be considered. When establishing the workforce mix of an activity, manpower planners shall review both peacetime and wartime missions so that activities are designed to transition easily from peacetime to wartime operations.

Military

DoDD 1100.4, *Guidance for Manpower Management*. Manpower in support activities shall not be designated as military solely for the purpose of exercising military authority under Sections 801-946 of 10USC.

DoDI 1100.22, *Guidance for Determining Workforce Mix*. Direction and Control of Combat and Crisis Situations--Command within the Armed Forces is implemented through a unique construct of command authority, known as the "military chain of command," and enforced, among other means, by the Uniform Code of Military Justice. Within the operating forces, this authority begins with field commanders and extends to the lowest level of command responsible for discretionary decision-making, personnel safety, and mission accomplishment. Accordingly, manpower in operational command of military forces is designated military.

Military personnel must obey all lawful orders at all times and are trained and prepared to immediately perform all duties as directed by military commanders. Military personnel may have to assume responsibility for discretionary decisions, to include assuming command and control of military operations, if the commander is killed or incapacitated during hostilities. Military personnel alone have this responsibility and military forces alone perform this role in defense of the nation. Combat authorized by the U.S. government is . . . designated for military performance. . . . Manpower that takes a direct part in hostilities shall be designated military if the planned use of disruptive and/or destructive combat capabilities (including offensive cyber operations, electronic attack, missile defense, and air defense) is an inherent part of the mission. Only military forces provide the appropriate authorities and controls (command authority, Uniform Code of Military Justice (UCMJ) authority, and discretionary decision authority), discipline, weapons, equipment, training and organization needed to execute combat missions on behalf of the United States. When the DOD authorizes the use of force to achieve a specific objective or end state, decisions on how to accomplish the mission are the responsibility of the Department. Security forces that operate as part of a larger, totally integrated and cohesive Armed Force typically perform operations that require deadly force and substantial discretion. These operations require military training and discipline and are designated for military performance. Security in hostile environments shall be designated for military performance if, in the judgment of the commander, the security operations would entail defense against military forces of another sovereign government or nonstate actors (e.g., terrorists or insurgents) whose capabilities are so unpredictable or sophisticated or such a high risk that a military capability is needed to provide the discretionary authority, flexibility, and options necessary to achieve the sovereign interests of the United States. Services provided by military medical personnel and Chaplains embedded in military units that engage in hostile action shall be designated military medical personnel function as an inherent part of the unit. Control of prosecutions and performance of adjudicatory functions in support of UCMJ criminal justice proceedings shall be designated military. Direction and control of intelligence interrogations, if performed in hostile areas where security necessary for DoD civilian performance cannot be provided, shall be designated military. Certain law enforcement operations, to include issuing warrants, making arrests, and preservation of crime scenes, that are carried out under the UCMJ, or are performed in hostile areas where security necessary for DoD civilian performance cannot be provided, shall be designated for military performance. Direction and control of detention facilities for Enemy Prisoner of War (EPW), Civilian Internee (CI), Retained Personnel (RP), other detainees, terrorists, and other criminals in areas of operation are IG activities . . . must be performed by military personnel.

Manpower authorities shall designate commercial Combat Service (CS) or Combat Service Support (CSS) functions in operating forces for military performance if, in the commander's judgment, a military capability is not normally required for proper performance of the duties, but performance of the function by DoD civilians or contractors or total reliance on DoD civilians or contractors would constitute an unacceptable risk. Manpower authorities shall establish sufficient manpower in the infrastructure so that an adequate pool of personnel is available for critical assignments in the operating forces during a mobilization, crisis, or war. Manpower in the infrastructure shall be designated military if the incumbents are Active military or Active Guard and Reserve (AGR) who are designated for assignments in the operating forces or serve as replacements for personnel in the operating forces during mobilization or war but perform CAs in the infrastructure during peacetime. Manpower authorities must consider the entire mobilization and wartime manpower demand before making final decisions about the numbers and skills required. Manpower performing DoD Functions shall be designated military if the incumbents are specifically required to provide "military advice" and counsel to the President, Congress, National Security Council, Secretary of Defense, senior DoD officials, or Secretaries. Commanders of the Combatant Commands are responsible to the President and Secretary of Defense for the execution of military missions, and require military judgment based on extensive military experience and cannot be transferred to DoD civilians or to the private sector. Manpower in infrastructure activities performing DoD functions shall be designated military if the incumbents have the authority to commit the Department of Defense to take action by direction, order, policy, regulation, contract, authorization or otherwise or have responsibility for activities, and the work requires military-unique knowledge and skills.

Officer

DoDI 1100.22, *Guidance for Determining Workforce Mix*. Only commanding officers and, in certain cases, commissioned officers have authority to order the arrest or confinement of members of the U.S. Armed Forces or civilians accompanying Armed Forces in the field during a declared war who have violated the UCMJ. The exercise of judicial and non-judicial punishment under the UCMJ is a responsibility assigned to military officers under sections 951 and 815 of Reference (c).

Enlisted

Not applicable.

Civilian

DoDD 1100.4, *Guidance for Manpower Management*. Manpower shall be designated as civilian except when military incumbency is required for reasons of law, command and control of crisis situations, combat readiness, or esprit de corps; when unusual working conditions are not conducive to civilian employment; or when military-unique knowledge and skills are required for successful performance of the duties.

DoDI 1100.22, *Guidance for Determining Workforce Mix.* Manpower shall be designated as civilian except when military incumbency is required by law, Executive Order, treaty or IA; for command and control of crisis situations, combat readiness, risk mitigation, or esprit de corps; when unusual working conditions or costs are not conducive to civilian employment; or when military-unique knowledge and skills are required for successful performance of the duties. In addition, sufficient military manpower shall be provided for overseas and sea-to-shore rotation, career progression, and wartime assignments.

Chairman of the Joint Chiefs of Staff

General Guidance

CJCSI 1001.01, *Joint Manpower and Personnel Program.* This instruction applies to all civilian positions, military positions graded O-6 and below, contractors and other assigned personnel (e.g., individual augmentees).

Grade. Technical, supervisory and experience requirements should determine the position's grade level.

(1) The position's grade must be consistent with the specific skill or specialty.

(2) The joint activity's aggregate grade structure should be consistent with mission requirements and Service grade ceilings.

Skill Code. Identify the job skill required for in the position using the military specialty code that best fits the position. This data will be listed in the "Skill 1" field (e.g., designator, rating, AFSC, military occupational specialty (MOS)). Identify the Navy officer billet classification code or additional skill identifier (ASI) in the "Skill 2" field when required. Identify the sub-specialty code, second ASI, or special qualification identifier in the "Skill 3" field when required.

Each January and July, the Services will submit to all JMP activities an updated list of skill codes that require special management, either because they have such low density or because the requirements exceed the inventory.

The manpower authorizations in the Joint Table of Distribution should be compared with Service manpower allocations shown in their FYDP exhibits by program year, PEC and manpower category (officer, enlisted, civilian). Any disconnects must be resolved.

Once a manpower authorization is approved and forwarded to the Services, the entire process usually takes 13 to 22 months from the date the manpower change request leaves the joint activity until the new individual reports for duty.

More coordination is required in the joint community than in the relatively streamlined world of a commander of a Service major command that allocates and manages a budget and endstrength to accomplish the assigned mission. Unlike Service major commands, joint activities do not own their manpower. Although the combatant commands and joint activities have been granted primary military missions by law, they have not been granted their own manpower budgets to manage. Combatant commands and joint activities control their own statements of requirements based on assigned missions but must coordinate with the Joint Staff, Services or resource sponsor on manpower funding and the Services with personnel inventory issues.

The Structure and Manpower Authorization System (SAMAS). SAMAS contains the authorizations at the military identity (officer, warrant officer and enlisted) and Army PEC level.

CJCSI 3500.01D, *Joint Training Policy and Guidance for the Armed Forces of the United States*. Individual joint training -- (Senior officer, senior enlisted, or individual staff officer.) Training that prepares individuals to perform duties in joint organizations (e.g., specific staff positions or functions) or to operate uniquely joint systems. Individual staff training ensures that individuals know, are proficient in, and have the joint competencies and skills to apply joint doctrine and procedures necessary to function as staff members. Individual joint training can be conducted by the Office of the Secretary of Defense, the Joint Staff, combatant commands, and activities and agencies responsive to the Chairman, Services, Reserve forces, or CSAs.

Military

CJCSI 1001.01, *Joint Manpower and Personnel Program*. Military positions should be established when required by law, training, security, discipline, rotation, combat readiness or when a military background is required for successful performance of the duties involved. Unusual working hours, difficult working conditions and avoiding overtime costs are not appropriate reasons for using military positions.

Military positions may not be used to replace civilian positions deleted by reduction in force programs. However, military personnel may be assigned temporarily to perform the functions of civilian positions when reduction in force programs cause temporary workload imbalances or mission impairment.

NATO grades are reflected by OF (officer) and OR (other ranks) category. NATO OF grades corresponds [sic] to US grades less one grade (e.g., US grade O-6 is NATO grade OF-5). NATO OR grades are equivalent to US enlisted grades (e.g., an OR-5 is an E-5).

New US-authorized manpower normally will not be approved without identification and deletion of an equal number of like-Service, like-grade offsets.

The staffing of a Security Assistance Organization (SAO) is the responsibility of the combatant command. Technical changes in job title, skill code, grade and specialty branch for positions other than the SAO Chief may be resolved between the combatant commands and the Chairman of the Joint Chiefs of Staff who will coordinate with the Services.

Combatant commands may recommend changes in grade or Service responsibility for an individual country using the criteria listed below:

a. Authorized personnel strength and composition.

b. Specific programs to be managed.

c. Host-nation forces and programs.

d. Relationship and attitude of the host-nation forces to US national objectives, strategic plans and military objectives.

e. Degree of importance and prestige the host-nation attributes to the components of its armed forces.

f. Geographic Service balance.

g. Grade required and justification.

h. Service, grade and proximity of other senior US military officials in country; and/or views of the US Chief of Diplomatic Mission.

Officer

CJCSI 1001.01, *Joint Manpower and Personnel Program*. The Officer Distribution Plan (ODP). Since there are more officer requirements than resources, the Army developed an ODP from the PMAD to distribute Active Component officer resources. The Army's joint manpower policy is to fill Joint Duty Allowance (JDAL) positions at 100 percent and the remainder of joint positions at the Army average for a particular skill and grade.

The Services should support all approved joint authorizations. However, mismatches sometimes occur between approved joint authorizations and the Services' existing personnel inventory to fill the authorizations. The mismatch is more significant between O3 and O4 officer positions since such a small proportion of other paygrades are found on staffs in the joint community. For those skills and grades that are critically short, the Services may coordinate a suitable substitution with the impacted joint activity and the Joint Staff/J-1.

A critical JDA position is one that, considering the duties and responsibilities of the position, requires a Joint Specialty Officer (JSO), an officer particularly trained in and oriented toward joint matters. In order to have a critical position filled by someone other than a JSO, SecDef must waive it.

At least 50 percent of each Military Department's JDAs must be filled by officers who have been designated as JSOs or JSO nominees.

JDAL Validation Board Procedures. All proposed JDAL changes will be routed through the Chairman of the Joint Chiefs of Staff who, after coordination with the Services, will provide a recommendation to the Secretary of Defense for approval. A formal board will be scheduled only if requested.

a. The Joint Staff/J-1 will provide a JDAL validation procedure package consisting of board procedures, review/validation package preparation, definitions, standards and criteria and a consideration checklist to the activity requesting board review.

b. Activities will then prepare and return the required information in the specified format. Packages will include organization/mission description or charter, a diagram of the organization's structure, the individual position description, a manning document reflecting the authorization and which shows the authorization does not reside within the incumbent's Military Department and extracts from performance reports of the incumbent or predecessors.

Although not included on the JDAL, officers assigned to the staff of a temporary Joint Task Force Headquarters (JTF HQs) involved in combat or combat-related operations, may receive joint credit after the officers serves [sic] 90 days.

Critical joint duty assignment billet. A joint duty assignment position for which, considering the duties and responsibilities of the position, it is highly important that the assigned officer be particularly trained in, and oriented toward, joint matters. Critical billets are selected by heads of joint organizations, approved by the Secretary of Defense and documented in the Joint Duty Assignment List. (Joint Pub 1-02)

CJCSI 1330.01D, *Assignment of Officers (O-6 and Below) to the Joint Staff*. As mandated by references, officers nominated for assignment to the Joint Staff will be among those considered to be the most outstanding of their Service.

An officer assigned to the Joint Staff may not be assigned more than 4 years without prior approval by the Secretary of Defense.

Critical Occupational Specialty (COS) – A military occupational specialty designated by the Principal Deputy Undersecretary of Defense Personnel and Readiness (PDUSD)(P&R) from among the combat arms in the Army or equivalent military specialties in the Navy, the

Air Force, and the Marine Corps where a severe shortage of trained officers in that skill exists. An early release may be requested from Joint Staff/J-1 for an officer with a COS serving his or her initial JDA.

Experience-based Joint Duty Assignment (E-JDA): Such other assignments and experiences that demonstrate an officer's mastery of knowledge, skills, and abilities in joint matters, as determined under such regulations and policy as prescribed by the Secretary of Defense. E-JDAs may be shorter in duration; therefore may be aggregated to achieve the equivalent of a full tour of duty in an S-JDA.

Joint Qualified Officer (JQO) – An officer designated by the Secretary of Defense, with the advice and assistance of the Chairman of the Joint Chiefs of Staff, who is educated and trained in joint matters and has completed the Level III requirements for JQO designation. Replaces the legacy term "Joint Specialty Officer (JSO)." An officer must be in the grade of 0-4 or above to be designated as a JQO.

Officers also may not be reassigned to the Joint Staff within 2 years of their latest Joint Staff tour without prior approval by the Secretary of Defense.

Requisitions for end-of-tour replacements normally will be submitted to the Services at least 12 months before the desired reporting date and for all others as soon as the requirement is identified. Normally, RC vacancies are automatically advertised for Service nomination processes[;] therefore requisitions are not required unless the requirement changes.

The proposed reporting date normally will provide a maximum 2-week overlap. The reporting date will be confirmed at the time of selection and may not be changed unless coordinated with the Director for Manpower and Personnel/J-1. Two-week overlap applies to Active Component (AC) support only and should not include clearing time.

The grades of officers requisitioned will be the minimum required for performance of the job.

Specific qualifications will be essential for the position relative to the Service-specific specialty requested.

An officer assigned to a non-critical S-JDA position on the JDAL may be moved to another non-critical S-JDA within the Joint Staff during their tour.

A Joint Staff tour for officers (O-6 and below) is 3 years. The Service may request that an officer be released from the Joint Staff before completing a full tour under certain circumstances.

CJCSI 1330.05, *Joint Officer Management Program Procedures.* Develop and implement policies and procedures to ensure that quality officers, who are adept at their Service core competencies, are assigned to the OSD, the Joint Staff (reference g), combatant command headquarters, and Defense agency headquarters.

The objective of the Joint Qualification System is to ensure a systematic, progressive, career-long development of officers in joint matters and to ensure that officers serving as G/FOs have the requisite experience and education to be highly proficient in joint matters, as directed in 10 USC chapter 38. The JQS will also allow officers to begin accruing points for their joint experiences at the earliest time in their commissioned careers.

Enlisted

CJCSI 1001.01, *Joint Manpower and Personnel Program.* There should be only one dedicated senior-enlisted advisor in each joint command. Other Services may be represented on an additional-duty basis.

Positions specifically excluded from the JDAL include officer positions of grade O-3 and below and all enlisted positions.

Military personnel authorized to assist G/FOs are called personal staff. Officer aides are included in the manpower authorization for joint and international activities. Enlisted aides are authorized by the Service of the G/FO concerned and are not reflected in the JTD.

CJCSI 1320.01D, *Assignment of Enlisted Personnel to the Joint Staff.* Only enlisted personnel with the highest personal and professional competence will be nominated and selected for assignment to the Joint Staff. A Joint Staff tour for enlisted personnel is normally 3, but not more than 4 years.

The reference designates that the Under Secretary of Defense for Personnel and Readiness will provide, among other things, policies and procedures for assigning Service members to the Joint Staff.

Enlisted personnel will be requisitioned for Joint Staff requirements in a manner that will provide approximately equal representation among the Military Departments.

The grades of enlisted personnel requisitioned will be the minimum required for performance of the job. Specific qualification requirements will be essential to the position and relative to the Service-specific specialty requested. All enlisted personnel assigned to the Joint Staff must satisfy the security requirements specified on the requisition.

Enlisted personnel with service issued interim Top Secret clearances with a closed National Agency Check are acceptable. If sensitive compartmented Information read-on is required, adjudicated TS clearance is mandatory.

e. Requisitions for end-of-tour replacements will be submitted to the Services approximately 12 months before the desired reporting date.

f. The Services will routinely submit the name of at least one qualified and available enlisted member for each requisition received. If the Services are unable to meet specified reporting dates or other requirements, a written request for relief is submitted to the Director for Manpower and Personnel/J-1 (DJ-1) as early as possible.

g. For all replacement actions, the Joint Staff has the right of refusal. The Joint Staff and [sic] will make a decision regarding the acceptability of the enlisted member within 15 working days of receiving the qualification record and advise the Service concerned.

h. An enlisted member submitted for assignment to the Joint Staff will not be available for nomination or assignment to another organization or duty unless the Joint Staff and the appropriate Service agree.

i. If an enlisted member is not accepted, the appropriate Service will be notified, verbally and in writing, with an explanation for the non-acceptance. The Service will respond promptly to meet the specified reporting date.

j. If an enlisted member is reassigned within the Joint Staff, the Joint Staff will provide to the affected Service the new military position description.

CJCSI 1805.01, *Enlisted Professional Military Education Policy.* The expansion of the joint operating environment to all levels of war necessitates the expansion of JPME to enlisted personnel. While not mandated by law (as is the case for officers), this policy is a recognition that operating in joint, interagency, multinational, and coalition warfighting organizations and staffs requires that joint learning objectives must be made available to all enlisted personnel.

Senior enlisted personnel (E-6/7 – E-9) should embrace the greatest responsibility for ensuring continued growth of themselves and others.

The enlisted Joint Professional Military Education (JPME) continuum is designed to progressively develop the knowledge, skills, perspectives, and values essential for enlisted personnel so they may function effectively in joint, interagency, and multinational operations and organizations.

A brief synopsis of the three Enlisted Joint Professional Military Education (EJPME) levels follows:

(1) Basic EJPME. The basic EJPME level addresses the progressive educational guidelines that should be completed by pay grade E-6.

(2) Career EJPME. This level addresses educational guidelines for SE personnel in grades E-6/7 and above.

(3) Senior EJPME. This level includes two CJCS-sponsored assignment-oriented educational opportunities beginning at the senior enlisted level:

(a) Senior EJPME Course. A stand-alone Web-based course that educates senior enlisted serving in or slated to serve in joint organizations).

(b) Command Senior Enlisted Leader (CSEL) KEYSTONE Course. Prepares CSELs for service in a flag officer joint headquarters and parallels the G/FO CAPSTONE Course.

Civilian

CJCSI 1001.01, *Joint Manpower and Personnel Program.* Civilian personnel will be used in positions that do not require military incumbents for reasons of law, training, security, discipline, rotation or combat readiness and do not require a military background for successful performance of the duties involved.

Civilian grades are determined by the local civilian personnel office after analysis of the specific position description. Either the determined grade or two zeros must be listed in this field.

U.S. Army

General Guidance

AR 71-32, *Force Development and Documentation—Consolidated Policies.* Manpower Requirements Criteria (MARC) - MARC are HQDA-approved standards to determine mission-essential wartime position requirements for combat support (CS) and combat service support (CSS) functions in TOEs. MARC are derived from detailed studies performed for the various CS and CSS functions.

Military

AR 611-1, *Military Occupational Classification Structure (MOCS) Development and Implementation.* The primary objectives of the MOCS are to provide occupational classification and structure guidance to standardize classification of positions and soldiers; describe and provide career progression paths for all soldiers to colonel (COL), chief warrant officer five (CW5) or sergeant major (SGM); prescribe grading guidance for all positions contained in requirements (Table of Organization and Equipment (TOE)) and authorization Modified TOE (MTOE), Table of Distribution and Allowances (TDA), Augmentation TDA (AUGTDA), mobilization TDA (MOBTDA), and Joint Tables of Allowances (JTA) documents.

Position classification structure

a. Position requirements are identified by an alpha-numeric code that identifies the occupational skills required to perform the principal duties of a position. These data elements for officers consist of grade, branch, Functional Area (FA), Area of Concentration (AOC), immaterial, Skill Identifier (SI) and Language Identification Codes (LIC) codes.

b. The nine characters of the position requirement code correspond to the nine-character field in TAADS--R document format with the column headings, MOS, ASI/LIC.

(1) The first three positions will be coded with the AOC associated with a branch or FA or with an immaterial position code that best defines the expertise needed.

(2) The fourth and fifth positions will normally be coded with a "00" indicating that no further expertise is needed. However, a branch, FA or immaterial code can be used, if necessary, to further define the position requirements within the following guidelines.

(a) If the first three positions are coded with an immaterial code, no additional code will be used.

(b) If the first three positions are coded with a branch AOC, a FA area code can be used.

(c) If the first three positions are coded with a FA AOC, then a branch AOC, combat immaterial or logistics immaterial code may be used.

(3) The remaining four positions will be used to identify further skill qualifications or language requirements when necessary.

c. Positions will be coded in accordance with the skills and expertise required. Personnel involved in position coding should have knowledge of the requirements of the positions, the skills and qualifications identified within the branch/FA AOC and those identifiable through skill and language identifiers. Over specialization/over documentation is discouraged. In identifying positions requiring multiple qualifications, careful consideration should always be given to the principal qualifications required. Branch AOCs may not be paired with other branch AOCs. As an exception, Military Intelligence Branch (35) AOC 35D and 35G may be used as a FA with AOC 15C only. Functional area AOCs may not be paired with other FA AOCs.

AR 71-32, *Force Development and Documentation—Consolidated Policies.* The Deputy Chief of Staff for Personnel (DCSPER) will formulate, coordinate, and enforce policies governing military position authorizations, classification, grading, and personnel management. Personnel proponents will determine personnel supportability and ensure that personnel requirements are included in the BOIP and supporting documents, and ensure that changes to personnel requirements (to include changes to the Army military occupational classification and structure (AR 611-1, AR 611-101, AR 611-112, and AR 611-201)) that impact on BOIP development have been identified, coordinated, and initiated per AR 611-1 for approval to coincide with approval of the BOIP.

Basis-of-Issue (BOIP) Plans

A BOIP is a requirements document that states the planned placement of quantities of new equipment and ASIOEP, as well as the reciprocal displacement of equipment and personnel. The BOIP process identifies mission essential wartime requirements for inclusion into organizations based on changes of doctrine, personnel, or materiel. Materiel developers use BOIPs as input for concept studies, life-cycle cost estimates, and trade-off analyses during the research and development process. MACOMs use BOIPs to plan the equipment, facilities, initial provisioning, and personnel required to support new or improved materiel systems.

BOIPs list 100 percent of the wartime requirements for TOEs, TDAs, JTAs, ADOPs, and TDA augmentation to mobilization TOE.

BOIPs describe in detail the new item, its capabilities, and where (in what organizations) it is to be used. BOIPs also identify the ASIOEP. BOIPs include personnel changes caused by the introduction of new items into the Army inventory and address the MOSs needed to operate and maintain the equipment. The BOIP process directs development of the BOIP feeder data (BOIPFD), Qualitative & Quantitative Personnel Requirements Information (QQPRI), and BOIP and related documents by the materiel, combat, doctrine, and training developers[.]

The TOE System

The TOE system governs the development and processing of TOEs from concept approval through publication and distribution.

The TOE is the end product document of the Army's combat development process. It merges, in one document, the results of the requirements determination process. This includes—

(1) Operational concepts.

(2) Life cycle system management model (LCSMM).

(3) ORD.

(4) The BOIP/QQPRI process.

(5) MARC studies.

(6) Other related documents and requirements determination systems.

(7) The Training and Doctrine Command Force Design Update TRADOC FDU process to access the affordability, validity, and impact of future organization.

TOEs are the primary basis for stating Army requirements. This document heavily impacts the budget, the training base, efficiency, operational readiness, and overall management of Army resources.

Military organizations prescribed in a TOE will contain only U.S. Army military positions.

Standards of grade (SG)

TOE positions will be graded in accordance with SG guidance provided in the AR 611-series. Proposed revision to this guidance will be submitted (AR 611-1) to the appropriate personnel proponent.

SG will be applied to level 1. The requirement for a position may be deleted for decrementing levels 2 and 3.

SG for a TOE being developed for new design or concepts must be documented per the AR 611 series. Requirements will be based on MMEWR policies. Requests for grade structure changes must be submitted to the appropriate personnel proponent for initiation of AR 611-1 action.

MARC

MARC supported positions are computed in accordance with the procedures in chapter 5 and the approved criteria published on the FMBB. Deviations from published MARC will be justified.

Staffing for continuous (multishift) operations (CONOPS)

Category I units (see AR 310-25 for unit category definitions) by doctrine must have sufficient requirements to allow CONOPS. Staffing will be based on CONOPS principles. MARC-derived staffing for CS and CSS functions within category I units may be organized for single or double shift operations, but will not exceed MARC allowances.

Multiple shift capabilities will be provided only in a category II and III TOE that require a 24-hour day wartime operational capability. Normally, no more than two shifts will be authorized for a 24-hour operation. Maintenance functions are staffed on the basis of maintenance MARC analysis.

Position justifications

All TOE position requirements not supported by MARC or doctrinal publications will be justified and included in the DTOE package.

Direct combat position code (DCPC)

All TOE positions will be coded in accordance with current regulatory guidance.

AR 611-1, *Military Occupational Classification Structure Development and Implementation.* MOCS requirements. Officer, warrant officer and enlisted occupational identifiers will be authorized normally when required to identify both soldiers and positions; however, in exceptional cases, an identifier may be authorized for personnel or position classification only.

Officer

AR 611-1, *Military Occupational Classification Structure (MOCS) Development and Implementation.* The term "officer" includes officer and warrant officer; officer in the grade of lieutenant or above; and warrant officer that means all warrant officers, with or without commission.

Officer, warrant officer and enlisted occupational identifiers will be authorized normally when required to identify both soldiers and positions; however, in exceptional cases, an identifier may be authorized for personnel or position classification only.

Positions will be coded in accordance with the skills and expertise required. Personnel involved in position coding should have knowledge of the requirements of the positions, the skills and qualifications identified within the branch/FA AOC and those identifiable through skill and language identifiers. Over specialization/over documentation is discouraged. In identifying positions requiring multiple qualifications, careful consideration should always be given to the principal qualifications required. Branch AOCs may not be paired with other branch AOCs. As an exception, Military Intelligence Branch (35) AOC 35D and 35G may be used as a FA with AOC 15C only. Functional area AOCs may not be paired with other FA AOCs.

Some positions are not related to any branch or FA and can be filled by officers from any branch or FA. Other positions are not related to a specific branch or FA but require experience in combat arms, logistics or personnel. These positions will be identified by a two digit code with the alpha designator "A" placed in the third position. The coding for these types of positions is in DA Pam 611-21, Part I.

e. In identifying positions which have been validated for advanced degrees by the Army Educational Requirements System, the branch/FA related to the educational discipline required should be identified as the principal skill required.

Military grades serve as both pay grades and grades of rank. Grade authorizations contained herein are based upon appropriate consideration of both to ensure--

(1) Grade appropriate to the amount and level of responsibility involved.

(2) Rank necessary for the amount and level of both responsibility and authority involved.

(3) Equitable enumeration for duties performed and qualifications required.

Grades for command and staff positions are provided in generic tables by command for each type position. Only principal positions are included. The grade standards listed in DA Pam 611-21 are provided to ensure equitable position grading in all requirement and authorization documents. Agencies responsible for the preparation, review or approval of requirements and authorization documents will adhere to these grade standards. Grades for positions not listed will be determined by comparing the primary duties and qualifications with those of the most nearly related positions contained in the tables. This comparison will be made in the terms of the appropriate factors presented in paragraph 4-29. Normally, the grade of officers within an organizational element and the heads of subordinate organizational elements will be at least one grade below that of the immediate supervisor. A position authorized a Colonel will not have immediate subordinates of equal grade. The grade of a civilian supervisor must also be taken into consideration. The following two exceptions will apply:

(1) Positions of Chief of Staff authorized as a Colonel may have direct subordinates of equal grade.

(2) Positions authorized a general officer but which are unsupported may be filled with a Colonel and may have direct subordinates of equal grade.

b. In determining grades for positions not included in the authorization tables in DA Pam 611-21, consideration will be given to the following:

(1) Organizational setting. The extent to which an incumbent accrues status and responsibility/authority from the organizational environment in which he must operate. Three subfactors are provided to assist in identifying and evaluating the relevant components of this factor.

(a) Organizational level. The level, in the Army's organizational hierarchy, of the headquarters in which the position under consideration is located. Consider levels from team, section or detachment to Joint Chief of Staff/DoD level.

(b) Magnitude of organizational responsibility. One relative measure of the total organizational responsibility is the number of officer-level position [sic] in the headquarters of the organization. Consider the number of officer-level positions (officer, warrant and professional civilian GS-9 or above) in the headquarters identified in subfactor (a) above.

(c) Level of position within the organization. The level of the position in the organizational hierarchy of the headquarters identified in subfactor (a) above (command section, principal or special staff, headquarters support or service element).

(2) Positional responsibility authority. The incumbent's share of his organization's total responsibility. Six subfactors are provided to assist in identifying and evaluating the relevant components of this factor.

(a) Type of position. Identify the basic function of the position (commander, executive officer, chief of a staff element, assistant staff officer, equipment or system operator) and the grade of the immediate supervisor. These are indicators of the functional scope and relative importance of the position.

(b) Magnitude of supervisory responsibility. One measure of the positional share of the total organizational responsibility is the number of officer level positions within the organizational element which is directed and controlled by the incumbent. Include all officer, warrant and professional civilians (GS-9 or above) in this total.

(c) Independence. Consider the nature of the controls over the position and the extent to which the incumbent is left to his own devices to achieve the desired results.

(d) Communication demands. Consider the extent to which the position requires skills in oral and written communications.

(e) Lateral points of contact. The magnitude of the coordination and nonsupervisory functions. Consider the organizational level and grade of person with whom contact is maintained. Organizational categories should include military, public or industrial and the new media. Disregard contact associated with commemorative or special events that occur on an infrequent or irregular basis.

(f) Auxiliary authority/responsibility. The extent to which the position requirements include significant, recurring additional duties, such as committee/study group participation, administration of military justice, personnel guidance and counseling programs, interior guard and preventive maintenance or medicine programs.

(3) Criticality to organizational mission. The extent to which success or failure in the management of assigned activities affects the organization as a whole. Two subfactors are provided to assist in identifying and evaluating the relevant components of this factor.

(a) Effects of errors. The many activities or groupings of activities in any organization can be graded in the order of their criticality. Consider the extent to which the incumbent's actions and decision [sic] affect the current operational effectiveness of the organization as a whole.

(b) Effect on future organizational effectiveness. The extent to which the position incumbent can affect future operational effectiveness. Consider the extent to which the position requirements include the development of concepts, plans, programs or procedures for future organizational activities. Implied in this subfactor is a consideration of the foresight, creativity and originality required.

(4) Skills and knowledge required. The level of skills and knowledge required for assignment to the position and attainment of a satisfactory level of performance by the end of a 30-day orientation period. Three subfactors are provided to assist in identifying and evaluating the relevant components of this factor.

(a) Formal education. Consider the level of formal educational development required to deal with the academic theories, facts and information that will be encountered while serving in the position.

(b) General military educational development. The range and substance of the overall military perspective which are required to deal with the military data, methods, theories and problems that will be encountered while serving in the position. Equate this to the career development training program of the military professions.

(c) Unique specialty training. The extent of the requirement for subject matter expertise which is acquired only through attendance at military or industrial training courses especially designed for the military function under consideration. Consider the length of the training course required.

(5) Grade balance. This factor is to be used to prevent grade gaps and ensure a progressive/proportionate distribution of grades with respect to superior and subordinate positions within the organization.

Warrant officers.

A warrant officer is defined as: "An officer appointed by warrant by the Secretary of the Army based upon a sound level of technical and tactical competence. The warrant officer is the highly specialized expert and trainer who, by gaining progressive levels of expertise and leadership, operates, maintains, administers and manages the Army's equipment, support activities or technical systems for an entire career."

The principles of warrant officer management are for use in determining whether certain officer level positions, per appropriate regulations, should be designated for warrant officer incumbency. Such positions are those that predominately involve the direct supervision of performance of technical operations, administration, supply and maintenance activities.

Positions that meet all or most of the following criteria will be considered for classification as warrant officer positions:

(1) Positions encompassing supervision of several enlisted or civilian technical skills related to a specific authorized warrant officer MOS (such as utilities operation and maintenance or telecommunications).

(2) Positions requiring continuous application of unique aptitudes, talents or abilities (such as those required for musical direction or aircraft piloting which have been designated as officer level positions by statute or regulation).

(3) Positions must be based on operational combat requirements and maintenance of combat readiness under conditions of combat.

(4) Positions that normally include responsibilities that enlisted personnel are prohibited by statute or regulation from performing.

(5) Positions must support an occupational field which is highly technical either due to orientation toward complex equipment and high density of multiple models or types of equipment or due to a management system which is technical and dependent on a number of subsystems (such as electrical, mechanical or administrative occupational fields).

(6) Positions must support a career field that requires a high degree of technical and tactical skill not readily available within the officer or enlisted structures, that requires repetitive assignments and that is related to an occupational field in which it is necessary for the Army to remain competitive with private sector incentives.

(7) Positions in a headquarters staff section supervising an activity that provides a service to the headquarters, subordinate units or personnel of the command (such as food services, personnel administration or medical care).

(8) Positions in Military Assistance Advisory Groups or missions requiring a technical adviser.

(9) Positions that normally do not require command of tactical units.

(10) Positions for which a valid organizational requirement exists/or is projected to exist.

(11) When a valid organizational requirement exists, assistant positions may be designated for warrant officer use if the position being assisted is also designated for warrant officer incumbency.

Officer level positions in an organizational element headed by an officer may be designated for warrant officer incumbency provided--

(a) They meet all or most of the provisions of (1) through (11), above.

(b) They are required for supervision of a separate activity or function within the organizational element headed by an officer.

(c) They are primarily technically oriented, but require tactical competence appropriate to the MOS authorized.

The following types of officer level positions are not authorized for designation as warrant officer positions:

(1) Company/troop/battery commander or executive officer of tactical organizations.

(2) Motor officer, if duties of the position involve more than maintenance (such as allocation of automotive assets).

(3) Positions which exercise broad planning and operational jurisdiction over subordinate operating elements.

d. An officer position, with a warrant officer incumbent because of circumstances discussed in DA Pam 611-21, will not be changed to a warrant officer position with a warrant officer MOS unless the position is definitely identifiable with warrant officer skills.

Enlisted

AR 611-1, *Military Occupational Classification Structure (MOCS) Development and Implementation.*

Grade standards for enlisted positions.

The SG tables will be used to establish enlisted grades in authorization documents. The average grade distribution matrix at figure 6-1 will be used in determining equitable distribution of grades (plus or minus two percent). Positions not similar to or shown in an existing SG tables [sic] will be evaluated using the factors of grade coding listed in paragraph 6-16 to determine appropriate grading. Grading a position not represented in a SG table is considered an exception and will be authorized only as indicated in paragraph 3-7, this regulation. If the position is a permanent requirement, an action must be initiated, through the appropriate personnel proponent, to establish a SG.

Grade authorizations listed are meant to ensure—

(1) Grade appropriate to the amount and level of responsibility involved.

(2) Grade necessary for the amount and level of responsibility involved.

(3) Equitable enumeration for duties performed and qualifications required.

b. In determining grade for positions not included in the SG tables, consideration will be given to the following:

(1) Similar organizations. Where a grade determination does not exist in the SG tables for the organization being considered, attempt to identify the position in the most nearly similar organization in terms of unit type, geographical location, standard reporting code and/or level below MACOM for which a grade determination exists.

(2) Requisite experience level. Consider the nature and extent of practical experience required in the position. Experience involves an extended application of learned skills and knowledge.

(3) Skill type. Consider the type of skill being employed. Operational skills will normally lie on scale including operations, combat employment, inspection, instruction, integration or evaluation of major systems.

(4) Skill level. Consider the level of skill required among the following:

(a) Basic. Employment of skill under supervision.

(b) Semi-skilled. Sufficient knowledge and competence to employ skills under minimum supervision.

(c) Skilled. Sufficient knowledge and competence to employ skills under any condition.

(d) Highly skilled. Requires top performance and demonstration of highest degree of applied knowledge.

(5) Criticality to organizational mission. The extent to which success or failure in the management of assigned activities affects the organization as a whole. Three subfactors are provided to assist in identifying and evaluating the relevant components of this factor.

(a) Effect of errors. Consider the extent to which the incumbent's actions and decisions affect the operational effectiveness of the organization.

(b) Uniqueness of skill. Consider the number of positions within the organization or its parent unit with the same or similar skills to those of the incumbent.

(c) Battlefield isolation. Consider the degree of geographic isolation under which the organization operates which may hinder or prohibit support from parent of sister organization.

(6) Skills and knowledge required. The level of skills and knowledge required for assignment to the position and attainment of a satisfactory level of performance by the end of a

30-day orientation period. Three subfactors are provided to assist in identifying and evaluating the relevant components of this factor.

(a) Formal civil education. Consider the level of formal civilian educational development required to deal with the academic theories, facts and information to be encountered.

(b) Military education. Consider the range and subsistence of the overall military perspective required and the military career development training program of the MOS or career field to be graded.

(c) Specialty functional training. Consider the extent of the requirement for subject matter expertise which is acquired only through attendance at military or industrial training courses especially designed for the military function under consideration. Length of training courses, career development availability for training and rate at which skills decay following training should be considered.

Enlisted personnel will be used in positions classified in their PMOS. When that is not possible, they will be assigned in their secondary MOS. If an individual is to be used in an additionally awarded MOS, the additional MOS should be related to the CMF of his PMOS.

Civilian

Not applicable.

U.S. Navy

General Guidance

OPNAVINST 1000.16K, *Navy Total Force Manpower Policies Procedures*. Manpower requirements define the number of personnel required to perform the Navy's work and deliver the specified capability. Each manpower requirement equates to a specific manpower space which is assigned qualifiers that define the duties, tasks, and functions to be performed and the specific skills and skill level required to perform the delineated functions.

In general, there are four types of manpower requirements:

(1) Fleet manpower requirements are determined by Navy Manpower Analysis Center (NAVMAC) and include ship, squadron and other deployable unit requirements.

(2) Shore manpower requirements at Navy commands are determined by the Manpower BSOs with Enterprise/Enabler endorsement and submitted to CNO (N1) for final approval.

(3) Individuals Account (IA) requirements account for personnel in student, trainee, transient or holdee (STTH) status as well as Midshipmen on active duty. Holdees include patients, prisoners and personnel in the process of being separated (separatees). IA requirements are determined by CNO (N12).

(4) Outside Navy requirements such as Combatant Commander (COCOM), defense agency and OSD billets are determined using other processes, and are reviewed by CNO (N12).

Manpower authorizations comprise the personnel entitlement of Navy commands to provide the required capabilities (for Fleet/operational units) or to perform assigned Missions, Functions and Tasks (shore activities).

The authorization will equal the requirement in quality (rating, designator, paygrade, etc) unless constrained by resources, CNO (N1) policy, or legal limitations such as Title 10 constraints on E8, E9, LCDR, CDR, CAPT and Flag billets.

Fleet manpower requirements shall be based on Required Operational Capability/Projected Operational Environment (ROC/POE) documents. Navy shore manpower requirements shall be based on directed Missions, Functions, and Tasks (MFTs). All requirements are to be reflected on the Activity Manpower Document/Activity Workforce Document (AMD/AWD).

Manpower requirements shall reflect the minimum quantity, calculated using the approved Navy Standard Work Week, and quality of manpower required for peacetime and wartime to effectively and efficiently accomplish the activity's mission.

The primary factors considered in developing Fleet manpower requirements are: (1) ROC/POE, (2) DCNO (N1) approved staffing standards, especially Navy Standard Workweek, (3) Warfare publications (CONOPS, NTTP, EDORM, NSTM, Combat systems doctrine, etc.), (4) Maintenance requirements, (5) Navy Maintenance and Material Management (3-M) Systems, (6) Navy Training System Requirements, Acquisition, Key Performance Parameters, and Training Requirements Program Planning Management (TRPPM) documents.

Manpower requirements are initially published by NAVMAC as draft FMDs, SMDs, SQMDs, and SEAOPDET manpower documents. TYCOMS, Warfighting Enterprises, Enablers and activities have an opportunity to review and submit a reclama. Community Managers will also have an opportunity to review the draft document and provide assessment of potential community health impacts.

Fleet manpower requirements are subject to change when any of the requirements drivers (e.g., ROC/POE, NTTP, NSWW, etc.) or equipment/hardware changes.

SMD/FMD manpower requirements are determined by, but are not limited to, the following development elements: ROC/POE parameters and analysis (wartime MFTs); directed manpower requirements (e.g., Master Chief Petty of the Command (MCPOC), safety, career counselors); Operational Manning (OM), also known as Watch stations (e.g., weapons control stations, repair parties, bridge, lookouts, etc.); Preventive Maintenance (PM) (e.g., scheduled maintenance, and Maintenance Index Page (MIP) cards); Corrective Maintenance (CM) (normally a ratio of PM) (NAVMAC may use validated CM workload if deemed appropriate; Facilities Maintenance (e.g., industrial standards and space layout); application of approved staffing standards (when applicable); workload measurement and analysis; utility tasking (e.g., Underway Replenishment (UNREP), Connected or Vertical Replenishment (CONREP/VERTREP), Flight Quarters (FQ), Sea and Anchor Detail, etc.); allowances (e.g., Productivity Allowance (PA), Production Delay (PO), Make Ready Put Away (MR/PA) time); development of officer requirements; and Warfighting Enterprise, TYCOM, Enabler, and Activity review of draft documents.

Shore manpower requirements will be based on valid, approved workload drivers. The primary driver of shore workload is the Mission, Functions and Tasks (MFT) statement of the activity.

The shore manpower requirements process accounts for far more manpower resources than any other process yet is less standardized and less well understood than the other processes. A rigorous and professional shore requirements process is critical to the cost effective use of Navy resources and benefits the manpower BSO by facilitating the allocation of manpower resources in a manner that best accomplishes the mission of the commander.

A manpower requirement becomes an authorization when it is supported by approved resources (funding) or meets other established criteria.

Funded positions represent the commitment of resources from appropriate resource sponsor for manpower. Funded positions do not guarantee a precise match of personnel. Statutory, fiscal, and inventory limitations may individually or collectively cause mismatches between funded positions and the actual inventory.

Because manpower authorizations form the basis of all Manpower, Personnel and Training and Education (MPTE) demand signals, all activities should endeavor to maintain a stable base of billet authorizations. Changes to the personnel inventory take much longer to effect than changes to authorizations so changes to military manpower authorizations should be limited and made as soon in advance as possible.

The Chairman, Joint Chiefs of Staff Instruction 1600.01 of April 30, 1998, contains the policies, administrative instructions, and responsibilities for determining the manpower requirements and for managing the manpower resources of all Joint activities, including Combatant Commands, Office of Secretary of Defense (OSD), Defense Agencies, Joint Chiefs of Staff (JCS), NATO, International commands, Naval Reactors and Outside DoD activities, such as White House, State Department, etc.

Navy cannot determine requirements for Joint activities.

There is no NAVMAC involvement in any of the Joint activity billet changes, unless they are executing a global change that would impact Joint billets as well as all others.

CNO (NI23) does not control budgets or funds, but is responsible for the proper programming of end strength to support validated joint requirements, both AC and RC, officer and enlisted.

U.S. Navy Regulations 1990, Precedence, Authority and Command 1001, Officers of the Naval Service. Officers of the United States naval service shall be known as officers in the line, officers in the staff corps, chief warrant officers and warrant officers.

2. officers in the line of the Navy include the following officers in the grade of ensign and above:

a. line officers not restricted in the performance of duty;

b. limited duty officers designated for duty in line technical fields; and

c. line officers restricted in the performance of duty designated for engineering duty, aeronautical engineering duty, and types of special duty which include cryptology, Intelligence, public affairs, and oceanography.

3. Officers in the staff cm-ps of the Navy include:

a. officers in the Medical, Supply, Chaplain, Civil Engineer, Judge Advocate General's, Dental, Medical Service and Nurse Corps, not restricted in the performance of duty within their respective COTPX and

b[.] officers in staff corps designated for limited duty within their respective corps.

4. In the Navy there are chief warrant officers and warrant officers. Chief warrant officers and warrant officers whose technical specialty is within the cognizance of a staff corps are classified as the staff corps. All other chief warrant officers and warrant officers are classed as in the line.

5. Officers of the Marine Corps of and above the grade of second Lieutenant are officers in the line and include:

a. officer[s] not restricted in the performance of duties and

b. officers designated for limited duty in appropriate technical fields.

6. Chief warrant officers and warrant officers of the Marine Corps and classed as in the line.

7. The term "line officer of the naval service" includes line officers of both the Navy and the Marine Corps.

8. Within the Manual for Courts-Martial, United States, 1984, and the Manual of the Judge Advocate General, the term "officer" includes a chief warrant officer, but does not include a warrant officer, unless the context indicates otherwise.

Military

OPNAVINST 1000.16K, *Navy Total Force Manpower Policies Procedures*. Military Requirements. A requirement or position is military if the successful performance of duties is required:

(1) for reasons of law, executive order, treaty, or international agreement; or (2) for command and control of crisis situations, combat readiness, risk mitigation, or esprit de corps; or (3) when unusual working conditions are not conducive to civil service employment; or (4) when military provide a more cost effective source of support; or (5) when military-unique knowledge and skills are required for successful performance of the duties. In addition, positions that do not meet the military essentiality test may be designated as military to: (1) provide overseas and sea-to-shore rotation; or (2) provide educational and career progression assignments; or (3) provide adequate military personnel to man wartime only assignments.

Military manpower authorizations generate demand signals to access, train, distribute and retain military personnel either directly or indirectly.

OPNAVINST 1001.24, *Individual Augmentation (IA) Policy and Procedures*. With or without the existence of a PRC authority, individual augmentation (IA) is not intended as recourse to increase peacetime manning levels, to meet training requirements or to fill permanent manning shortfalls. CNO will support fully validated manning requirements identified by combatant commands and their associated Navy component commands.

Officer

NAVPERS 158391, *Manual of Navy Officer Manpower and Personnel Classifications*.

Officer Billet Designator Codes

The officer billet designator codes are four digit numbers used to identify the primary naval specialty qualifications required of the billet incumbent and to administratively categorize officer billets for proper management and identification. They serve as a manpower management tool when used in conjunction with the officer designator codes. These codes are entered in the Manpower Authorizations (OPNAV Form 1000/2) to indicate the categories of officers required for the billets.

Officer Designator Codes

The officer designator codes are four digit numbers used to group officers by categories for personnel accounting and administrative purposes and to identify the status of officers. These codes identify, through the first three digits, the categories in which officers are appointed and/or designated and, through the fourth digit, the status of the officers within the various categories. A listing by the first three digits of the officer designator codes and their description and the translation of the fourth digit are included in the following section.

Navy Officer Billet Classifications (NOBCs) identify officer billet requirements and officer occupational experience acquired through billet experience or through a combination of education and experience. An NOBC provides a general description of duties performed in a billet. It is not meant to cover every duty required of the job nor is every duty listed necessarily a requirement of the billet. Similarly, an NOBC in an officer's record does not necessarily indicate that the officer has experience in every duty listed in the definition. The NOBC identifies a group of officer billets which are similar but not necessarily identical in scope and nature of duties.

When applied to a billet, the NOBC is a general statement of duties performed in accomplishing some part of an activity's mission.

Assignment of an NOBC on all officer billets is mandatory.

All changes to manpower authorizations must result in a billet base that is executable in terms of resulting in systemically sustainable communities that are achievable within fiscal controls (including special and incentive pays) and legal limits on control grades.

The Officer Subspecialty System is an integrated manpower and personnel classification and control system that establishes criteria and procedures for identifying officer requirements for advanced education, functional training, and significant experience in various fields and disciplines. The Subspecialty System is used to identify those requirements which require a Subspecialty Code (SSC) including a specific degree program (four digit code) and the specific level (suffix last digit) required to perform the task and mission of the validated position. Subspecialty needs will be validated for the minimum education level deemed essential for performance. The term "level" does not necessarily imply the need for a degree, but rather "level" indicates the minimum required education and/or training and/or experience. Undergraduate education majors, specialized functional training programs, and significant experience will also be used to meet subspecialty needs. Graduate education supports the officer subspecialty system and is designed to complement the officer's operational training.

OPNAVINST 1210.2B, *Navy Officer Occupational Classification System (NOOCS).* NOOCS is the method the Navy uses to identify skills, education, training, experience, and capabilities related to both officer personnel and manpower requirements. This system forms the basis for officer personnel planning, manpower management, procurement, training, promotion, distribution, career development and mobilization.

The Designator/Grade structure consists of designators and paygrades that provide a framework for officer career development and promotion. This structure is the primary administrative means for classifying, identifying, and documenting officer manpower resources and requirements.

Under regulations prescribed by the Secretary of Defense, the Secretary of each military department shall establish competitive categories for promotion. The Navy uses officer personnel designators for this purpose.

The ACNP for Total Force Programming and Manpower (Pers-5) programs, coordinates, and reviews officer billet quality control and develops the officer qualitative and quantitative manpower authorizations.

The Designator/grade Structure. Since changes to designators and/or grades affect the basic identification of officers and billets, each change requires detailed analysis to evaluate the impact on manpower, personnel, and training matters. Members shall review all change requests in depth and make appropriate recommendations concerning proposals that add, delete, or revise designators and/or grades.

Navy Regulations 1001, Officers of the Naval Service. 1056. Command of a Ship.

1. The officer detailed to command a commissioned ship shall be an officer of the line in the Navy eligible for command at sea.

2. The officer detailed to command an aircraft carrier, an aircraft tender, or a ship with a primary task of operating or supporting aircraft shall be an officer of the line in the navy, eligible for command at sea, designated as a naval aviator or naval flight officer.

1059. Command of a Staff Corps Activity. Officers in a staff corps shall be detailed to command only such activities as are appropriate to their corps.

SECNAVINST 1000.7F, *Interservice Transfer of Commissioned Officers*. Interservice Transfer. The transfer of commissioned officers serving on active duty, between uniformed services, and the transfer of commissioned officers not on active duty, between the Reserve components of the uniformed services.

Commissioned officers qualified to contribute to an activity of another uniformed service will be given an opportunity to apply for interservice transfer without interruption in their service career.

a. Interservice transfers are intended for use primarily in technical fields to permit full use of specialists, but will not be restricted to technical specialists. Interservice transfers may be used to help fulfill authorized strength requirements in competitive categories, designators, occupational fields, military occupational specialties, and other authorized officer specialties. They may not be made from shortage specialties.

Eligibility for Transfer to the Navy or Marine Corps. All officers of other services are eligible for transfer to the Navy or Marine Corps except those who:

a. Have been deferred from promotion or have failed selection for promotion one or more times in present grade.

b. Are in year groups that, in the candidate Navy or Marine Corps competitive category, designator or Military Occupational Specialty (MOS), are filled.

c. Have applied for transfer to the gaining service within the last year.

d. Have been notified of mandatory retirement for any reason.

e. Are inactive duty Reservists with a remaining Military Service Obligation (MSO) unless:

(1) The officer has or is willing and able to acquire special experience or professional, educational, or technical skills of greater value to the gaining component than to the losing component.

(2) The officer has skills that exceed the requirements of the parent component and are needed in the gaining component in an approved competitive category, designator or MOS which is below authorized strength.

(3) For SELRES the losing component has no organized paid-drill unit to which the member could be usefully assigned within commuting distance of the members present or future home or place of business, and there is a Navy Reserve of Marine Corps Reserve activity within commuting distance with a desired billet available.

Enlisted

NAVPERS 18068F, *Manual of Navy Enlisted Manpower and Personnel Classifications and Occupational Standards*, Vol. II: *Navy Enlisted Classifications (NECs)*. The Navy Enlisted Classification (NEC) system, of which the NEC coding system is a part, supplements the enlisted rating structure in identifying personnel on active or inactive duty and billets in manpower authorizations. NEC codes identify a non-rating wide skill, knowledge, aptitude, or qualification that must be documented to identify both people and billets for management purposes. Additionally, an NEC can be used to identify special circumstances or situations with approval via the NEOCS process.

The NEC coding system facilitates management control over enlisted skills by identifying billets and personnel and enhancing efficient use of personnel in distribution and detailing. In cases where NECs reflect special training, inventories of coded billets and coded personnel are also the basis for planning and controlling input of personnel into formal courses that earn NECs. Consequently, the continuing enlisted strength of the Navy, particularly petty officer allocations, and funds authorized for rating and specialty training depends, to an increasing extent, upon the accuracy, thoroughness, and timeliness of NEC coding. Personnel required to support ratings and special programs must be identified by the correct combinations of rates and/or NECs.

Personnel are automatically assigned NECs upon satisfactory completion of applicable courses.

OPNAVINST 1000.16K, *Navy Total Force Manpower Policies Procedures*. Enlisted manpower requirements are considered general duty when multiple rates or ratings could satisfy the work requirement (e.g., Recruiters, Recruit Division Commanders, NAVLEAD Instructors, 3M Coordinator, etc.) either because the work is not directly associated with any one rating (codified using FAC G) or because the work is associated with a multisourced NEC (codified using FAC H). To provide maximum flexibility in balancing sea shore rotation and career progression, enlisted requirements will be classified as general duty whenever appropriate.

The need for [enlisted] priority manning is dictated by the fact that mission accomplishment of some activities is especially essential to the national interest and these activities must be properly manned, even when personnel shortages exist. Priority manning may be on a continuous basis or may exist only for a specified period of time. Continuous management of authorized priority manning is necessary to ensure mission accomplishment and to provide the order of importance for which activities will be priority manned while undermanning other activities. The Manning Control Authorities will establish manning levels within their area of responsibility to the authorized priority manning requirements in order of Priority (1, 2, & 3), and then generally provide (fair share) manning levels for their remaining activities.

OPNAVINST 1223.IB, *Navy Enlisted Occupational Classification System (NEOCS)*. NEOCS is the method the Navy uses to identify enlisted personnel skills and the requirements associated with these skills. The system forms the basis for actions taken concerning enlisted personnel planning, manpower management, procurement, training, promotion, distribution, assignment, and mobilization.

Pers-2 develops and monitors policies concerning military personnel programs dealing with enlisted professional development, use, and overall career force maintenance.

Resource sponsors assist N1 and CHNAVPERS in executing their responsibilities with regard to enlisted manpower and personnel and have overall technical cognizance and resource responsibility for the respective warfare/specialty area.

Civilian

OPNAVINST 1000.16K, *Navy Total Force Manpower Policies Procedures*. Total Force requirements shall reflect the appropriate mix of military, civil service and private sector manpower necessary to accomplish Department of Defense (DOD) missions consistent with applicable laws, policies, and regulations.

Non-military manpower requirements which consist of functions and tasks that are inherently governmental (IG) in nature shall be performed by government civilians. IG functions may include activities that require either the exercise of discretion when applying federal government authority or value judgments when making decisions for the federal government. IG requirements must be filled with military or civil service personnel. In addition, a nonmilitary essential requirement is identified as civil service if incumbency is required due to:

(1) law, executive order, treaty, or international agreement; or

(2) for key personnel and emergency essential civil service manpower; (see Section 407 for complete definition); or

(3) for continuity of infrastructure operations during national emergency or war; or

(4) core logistics capability.

Manpower requirements that do not meet the criteria for military or government civilian requirements shall be designated as contractor requirements unless military or civilian manpower can be demonstrated to be more cost effective.

The Navy relies on its civilian work force to support military forces and systems in peacetime. The need for stability in highly technical and hard-to-fill positions has resulted in civilians occupying positions that provide a function or service that would be essential to support of combat forces during periods of natural disaster, rising tensions, or actual hostilities, including a mobilization. These positions are designated Key and/or Emergency-Essential (E-E).

U.S. Air Force

General Guidance

AFPD 38-2, *Manpower and Organization.* Manpower Authorization—A funded manpower requirement with detail that defines the position in terms of its function, organization, location, skill, grade, and other appropriate characteristics that commands use to extend manpower end strength to units.

Manpower Requirement—A statement of the manpower needed to accomplish a specified job, workload, mission, or program. There are two types of manpower requirements: funded and unfunded. Funded manpower requirements are validated and allocated manpower needs. Unfunded requirements are validated manpower needs but are deferred because of budgetary constraints.

AFI 38-204, *Programming USAF Manpower.* The Air Force uses three types of manpower to perform required work:

Military personnel.

In-service civilian employees.

Contract services.

AFI 36-2101, *Military Personnel.* The objective of the military personnel classification system is to identify duties and tasks for every position needed to accomplish the Air Force mission. The system is designed to identify qualifications and abilities necessary to accomplish these duties and tasks, as well as provide clear and visible career progression patterns. It links duties and tasks into cohesive job clusters that are used to match personnel requirements with personal aptitudes, attributes, and qualifications. The classification system also provides concise award, upgrade, and retention criteria for career progression.

AFPD 38-2, *Manpower and Organization: Manpower*. To support national security objectives, Air Force units must successfully accomplish assigned missions using minimum levels of manpower needed to effectively and efficiently execute missions. Air Force provides unit commanders, through the commands, with manpower in quantity, grade, and specialty required to accomplish assigned missions

AFI 38-201, *Manpower Determining Manpower Requirements*. The Air Force manpower requirements determination process systematically identifies minimum essential manpower required for the most effective and economical accomplishment of approved missions and functions within organizational and resource constraints.

Air Force Manpower Standard (AFMS) development results from reviewing all available options for mission accomplishment. Determining whether the function is inherently governmental, in whole or in part, is a first step. Second, carefully scrutinize each function and decide if it lends itself to competitive sourcing (i.e., competitive sourcing, OMB Circular A-76 cost comparison, privatization, etc.). Determine which component (active, AGR) is most appropriate when active duty resources are not required. Finally, organizationally reengineer the function including best practices and advanced technologies, etc. The reengineering results are then rolled into an existing AFMS, become a new AFMS, or manpower determinant.

3.1. Why We Allocate Air Force Military Grades:

3.1.1. Military positions required to accomplish Air Force missions are identified by grade and skill using the manpower determination process.

3.1.2. Congressional ceilings constrain the number of active duty general officers, field grade officers, and noncommissioned officers in the grades of senior master sergeant and chief master sergeant. The DoD Authorization Bill defines the maximum number of officers and enlisted authorizations.

3.1.3. The Air Force considers enlisted skill level, rather than grade, as the prime indicator of what is required to do the job. Therefore, the integrity of stated skill level requirements must be maintained when implementing enlisted grade adjustments.

3.1.4. The Manpower Data System (MDS) documents required and authorized grades for all military manpower requirements. The required grade data field of the MDS states unconstrained required grades to accomplish specific workloads. Authorized grades in the MDS reflect fiscal reality and define grades allowed by applying allocated grade factors to the budgeted end strength.

3.2. Responsibilities for Grades Allocation:

3.2.1. Air Force Director of Manpower, Organization and Quality (HQ USAF/XPM):

Provides overall planning and policy guidance.

Approves fixed grade allocations.

Approves officer and enlisted grade allocations.

3.2.2. Manpower Requirements Division (HQ USAF/XPMR):

Obtains budgeted end strengths, statutory grade ceilings, and general officer positions.

Validates colonel positions.

Identifies fixed positions.

Assists functional OPRs with enlisted career progression group (CPG) review.

Provides final grade factors to MAJCOMs, FOAs, and DRUs.

Monitors commands for compliance with grade factors.

3.2.3. Air Force Center for Quality and Management Innovation (AFCQMI):

Defines and describes data system requirements to support the grades allocation process.

Provides systems analysis, programming support, and data processing support for developing and maintaining grades management computer programs.

Analyzes and evaluates results.

Prepares final grade factors for command allocations.

3.2.4. Air Force Personnel Center (AFPC). Provides historical student, transient, and personnel holdee data by grade for officers and enlisted.

3.2.5. Air Staff Functional Managers. Review enlisted CPG allocations, prior to development of final grade factors, for unique requirements and self-sustaining structures.

3.2.6. MAJCOMs, FOAs, and DRUs. Ensure required grades data in the MDS reflects accurate command grade requirements, and implement new fiscal year grade allocations within time constraints levied by HQ USAF/XP. Command totals will not exceed grade levels allocated.

3.3. Grades Allocation Process. The following procedures are used to allocate military grades to MAJCOMs, FOAs, and DRUs for active duty military forces (in File Part A of MDS) on a biennial basis.

3.3.1. Determine Air Force Grade Ceilings. For officers, use the Defense Officer Personnel Management Act (DOPMA) to determine the number of colonel, lieutenant colonel, and major grades. Extrapolate ceilings for field grades, based on funded officer end strength for the allocation years, from the appropriate table in DOPMA. Because there is no specific ceiling outlined in DOPMA for company grade officers, captain and lieutenant grade allocations approximate the projected personnel inventory for the allocation years. By law, chief master sergeant and senior master sergeant grades are constrained to three percent of the total budgeted end strength for the enlisted force, of which no more than one percent may be chief master sergeant. Apply these percentages to funded enlisted end strength for the allocation years. Allocate the remaining enlisted grades based on the Air Force Enlisted Grades Plan.

Exclude the following from the process:

Active Guard/Reserve Positions (Functional Categories X, U, V, Y)

Medical Corps (MC) and Dental Corps (DC) Positions (AFSCs 44XX, 45XX, 47XX, 40C0C, 40C0D, and 48XX)

3.3.2. Subtract Student/Transient/Personnel Holdee (Patients, Prisoners, and Separatees) grades from Air Force grade ceilings. HQ USAF withholds and does not allocate nonpermanent party students (carried in File Part D of MDS), transients, and personnel holdees to the commands. The Air Force Personnel Center (AFPC) provides historical data for withheld grades.

3.3.3. Subtract fixed grades, approved and provided by HQ USAF/XPM, from the total grades available.

The following categories are fixed at the authorized grade in the MDS:

General Officer positions.

Colonel positions.

Outside the Air Force positions.

Reimbursable positions.

Miscellaneous fixed positions (functions or units with unique mission requirements approved by HQ USAF/XPMR).

3.3.4. Distribute remaining grades. Use a "fair share" percentage of funded grade requirements.

3.3.5. Allocate Grade Factors to HQ USAF, MAJCOMs, FOAs, and DRUs. HQ USAF/XP provides command grade factors derived from the sum of fixed and fair share grades for all officer and enlisted positions. Grades for enlisted positions are allocated by Career Progression Group (first three digits of the Air Force Specialty Code).

3.4. Grade Changes. Submit requests for changes in grade factors to HQ USAF/XPMR

Responsibilities and Guidelines for AGR Positions:

7.1.1. MAJCOMs establish AGR authorizations for members of the Air National Guard of the United States (ANG) or Air Force Reserve Command (AFRC) voluntarily serving on extended active duty or active duty (other than training) for a specified time, usually over 360 days. Do not establish authorizations to circumvent Congressional or Secretarial limitations on management headquarters activities or Air Force personnel strengths. AGR authorizations are subject to the usual budgetary review just as any other Air Force program.

Military

AFI 38-204, *Use of Military Personnel.* In accordance with DoD guidance, the Services assign military personnel only to positions that:

Directly contribute to prosecution of war (combat or direct combat support).

Are military by law.

Are military by custom or tradition.

Are needed for overseas rotation, operational deployment augmentation, career field sustainment.

Manpower Mix. Military essentiality is the determinant of the Air Force manpower mix.

Military Work. When currently authorized civilian positions are identified as military essential, commands should convert the civilian authorization to military through attrition unless mission requirements dictate a faster conversion. Coordinate all conversions in functions affecting Air National Guard or Air Force Reserve forces training with HQ Air Force Reserve Center or the National Guard Bureau.

Designates as military essential those positions that:

Perform combat or direct combat support functions.

Require previous military training.

Are dictated by tradition or custom.

Require current military experience.

Require incumbents to exercise direct command authority over military subordinates under the Uniform Code of Military Justice.

The law requires.

Are indirect combat support positions tasked by a HQ USAF-approved contingency or war plan.

Require skills not available in the civilian community.

How to Determine Military Essentiality.

1 Requires incumbent to perform direct combat as part of a unit's mission (examples: fighter and bomber air crews, aerial gunners)

2 Requires incumbent to support direct combat (examples: aircraft and aircraft systems maintenance of combat coded aircraft, missile systems maintenance, munitions maintenance)

3 Deploys for combat or direct combat support augmentation to support AF-wide mission requirements (example: Air Education and Training Command security police deploying to a forward operating location)

4 Is required by law to be military (example: chaplains; per 10 U.S.C.)

5 Requires the exercise of direct military authority or discipline (example: commanders and first sergeants)

6 Is of a military nature supporting organizations outside the Air Force (example: JCS, FEMA)

7 Requires previous military training (example: air weapons director)

8 Is military by tradition or custom (example: bandsmen, USAF Honor Guard)

9 Supports indirect combat when use of volunteer civilian employees is deemed unreasonable (e.g.; information manager at an operating location) or is tasked against a UTC to support an OPlan providing probable exposure to hostile fire

10 Requires current military experience (example: basic military training instructors)

11 Is one for which no civilian skill or resource is currently available (example: data processing at remote sites)

Military Advanced Academic Degree (AAD) Requirements

8.1. Responsibilities For Developing And Reviewing AAD Positions:

8.1.1. HQ USAF functional managers or academic specialty monitors (ASM) develop criteria for assessing AAD requirements for their career field and will provide this criteria to MAJCOM, FOA, DRU, and joint activity functional managers and DP for use in the verification process. Annually, Asks attend the Air Force Education Requirements Boards (Avers) which verify graduate education requirements.

Verification Process. The MAJCOM, FOA, and DRU Exams will:

8.2.1. Identify all authorizations in the MDS currently possessing an Academic Specialty Code (ASC).

8.2.2. Provide a list of AAD coded positions to MAJCOM/DP representative for their use in directing the verification process. Each functional OPR or ASM reviews their AAD billets and authenticates them for mission essentiality.

8.3. Follow-On Action for Exams. After AFERB completion, a joint letter identifying approved quotas for coded positions will be sent to MAJCOM DPs/XPMs.

8.3.1. Positions identified for masters remain coded for 6.5 years and doctorate for 8 years unless a waiver is approved.

8.3.2. Out of cycle requirements, those identified prior to the yearly AFERB, are coded in MDS only after MAJCOM Functional OPRs had substantiated their need through their MAJCOM DP. These are restricted to mission essential.

9.2.3. FOAs, DRUs, and Joint Agencies.

9.2.3.1. Assign rated AFSCs only to those positions with duties clearly requiring rated skills as [d]escribed in AFM 36-2105. Ensure rated officer positions are established at the lowest organizational level consistent with mission requirements.

MILITARY GRADES

3.1. Why We Allocate Air Force Military Grades:

3.1.1. Military positions required to accomplish Air Force missions are identified by grade and skill using the manpower determination process.

3.1.2. Congressional ceilings constrain the number of active duty general officers, field grade officers, and noncommissioned officers in the grades of senior master sergeant and chief master sergeant. The DoD Authorization Bill defines the maximum number of officers and enlisted authorizations.

3.1.3. The Air Force considers enlisted skill level, rather than grade, as the prime indicator of what is required to do the job. Therefore, the integrity of stated skill level requirements must be maintained when implementing enlisted grade adjustments.

3.1.4. The Manpower Data System (MDS) documents required and authorized grades for all military manpower requirements. The required grade data field of the MDS states unconstrained required grades to accomplish specific workloads. Authorized grades in the MDS reflect fiscal reality and define grades allowed by applying allocated grade factors to the budgeted end strength.

Officer

AFI 38-201, *Manpower and Organization: Determining Manpower Requirements.* Colonel Positions. Colonel positions are a critical resource and are managed within congressional ceiling constraints.

4.3. Criteria for Core Colonel Positions. Use the following criteria to assess core colonel requirements at varying levels of organization. Variances to the criteria are considered on a case-by-case basis and approved by HQ USAF.

4.3.1. General Criteria. The following criteria apply to all organizations:

No colonel deputies.

No colonel positions more than one organizational level below another colonel position.

No colonel positions in same organizational structure (work center) as GS-15.

An Air Force Form 81 is not required for Objective Wing Commanders, Vice Commanders and Group Commanders. However, MAJCOMs may choose to maintain PDs for these positions.

4.3.2. HQ USAF and MAJCOM Headquarters. Colonels allowed only down to division level. MAJCOM/SGs will have at minimum the BSC, MSC, and NC represented.

Field Operating Agencies (FOA) and Direct Reporting Units (DRU). No more than two colonels in the headquarters staff. FOA and DRU subordinate units fall under appropriate criteria, such as, wing, group, center, or the like.

4.3.4. Numbered Air Forces. Colonels allowed only in following positions: Vice Commander, Operations Division, Logistics Division, and the Staff Judge Advocate.

4.3.5. Installation Level:

Objective Wing. Colonels allowed only in the following positions: Commander (if no general officer), Vice Commander, and Group Commanders.

Centers. Only down to division level.

4.3.6. Medical:

Installation Medical Treatment Facilities. Medical Group Commanders, Senior BSC, MSC, and NC at Level 10 medical treatment facilities.

Medical Centers. Colonels allowed only down to division level.

4.3.7. AFMC Criteria. The following criteria apply to AFMC only:

MAJCOM Headquarters, down to division level.

ALCs, Product Centers, System Program Offices (SPOs), Laboratories, System Program Directorates (SPD), USAF School for Aerospace Medicine, Medical Centers, down to division level.

Staff Judge Advocate positions in organizations where the Commander has General Courts Martial (GCM) convening authority.

Objective Wing Commander (if no general officer), Vice Commander, and Group Commanders.

Senior BSC, MSC, and NC at Level 10 medical treatment facilities.

No colonel deputies.

No colonel positions more than one organizational level below another colonel position.

Establishing New Colonel Requirements.

4.4.1. Core Positions. Commands can establish colonel positions that meet the criteria in Paragraph 4.3 without Air Staff approval, providing the command resources the colonel grade with an existing colonel grade (zero balance action).

Grades Allocation Process. The following procedures are used to allocate military grades to Majors, Foes and Drugs for active duty military forces (in File Part A of MDS) on a biennial basis.

3.3.1. Determine Air Force Grade Ceilings. For officers, use the Defense Officer Personnel Management Act (DOPMA) to determine the number of colonel, lieutenant colonel, and major grades.

Extrapolate ceilings for field grades, based on funded officer end strength for the allocation years, from the appropriate table in DOPMA. Because there is no specific ceiling outlined in DOPMA for company grade officers, captain and lieutenant grade allocations approximate the projected personnel inventory for the allocation years. By law, chief master sergeant and senior master sergeant grades are constrained to three percent of the total budgeted end strength for the enlisted force, of which no more than one percent may be chief master sergeant. Apply these percentages to funded enlisted end strength for the allocation years. Allocate the remaining enlisted grades based on the Air Force Enlisted Grades Plan.

Exclude the following from the process:

Active Guard/Reserve Positions (Functional Categories X, U, V, Y)

Medical Corps (MC) and Dental Corps (DC) Positions (Asks 44XX, 45XX, 47XX, 40C0C, 40C0D, and 48XX)

3.3.2. Subtract Student/Transient/Personnel Holder (Patients, Prisoners, and Separates) grades from Air Force grade ceilings. HQ USAF withholds and does not allocate nonpermanent party students (carried in File Part D of MDS), transients, and personnel holders to the commands. The Air Force Personnel Center (AFPC) provides historical data for withheld grades.

3.3.3. Subtract fixed grades, approved and provided by HQ USAF/XPM, from the total grades available.

The following categories are fixed at the authorized grade in the MDS:

General Officer positions.

Colonel positions.

Outside the Air Force positions.

Reimbursable positions.

Miscellaneous fixed positions (functions or units with unique mission requirements approved by HQ USAF/XPMR).

3.3.4. Distribute remaining grades. Use a "fair share" percentage of funded grade requirements.

3.3.5. Allocate Grade Factors to HQ USAF, Majors, Foes, and Drugs. HQ USAF/XP provides command grade factors derived from the sum of fixed and fair share grades for all officer and enlisted positions. Grades for enlisted positions are allocated by Career Progression Group (first three digits of the Air Force Specialty Code).

3.4. Grade Changes. Submit requests for changes in grade factors to HQ USAF/XPMR.

Enlisted

AFM 36-2108, *Enlisted Classification.* The occupational structure is flexible to permit enlisted personnel to specialize and develop their skills and abilities while allowing the Air Force to meet changing mission requirements. Individual enlisted personnel have a joint responsibility with commanders and supervisors at all levels to fully develop their abilities consistent with Air Force needs and within the established patterns of specialization.

Specialty descriptions and codes identify different types of Air Force jobs and the qualifications of enlisted personnel to fill these jobs. Individual specialty descriptions . . . provide the occupational standards required or desired for successful performance in an Air Force Specialty. Standards are used to procure, classify, and employ enlisted personnel; to develop career programs for initial training, retraining, and skill upgrade; and to structure unit manpower documents (UMD) positions.

Civilian

AFI 38-201, *Manpower and Organization: Determining Manpower Requirements.* Civilian Position Management. The Air Force's objective is to manage civilian human resources within Congressional constraints and the Air Force civilian pay budget while balancing mission needs with economy and efficiency of operations.

U.S. Marine Corps

General Guidance

MCO P1300.8, *Marine Corps Personnel Assignment Policy.* Promotion through the grade of lieutenant colonel for officers and master sergeant for enlisted (but not first sergeant) shall not be the sole reason for PCS transfer prior to tour completion[.]

MCO P1000.6G, *Assignment, Classification, and Travel System Manual.* Classification of military qualifications consists of assigning, changing, voiding, or converting MOS's to identify accurately each individual's current qualifications.

3203. TYPES OF BILLETS

1. Billets are of three types -- "A" billets, "B" billets, and billet designators.

a. "A" billets are identified by a category "A" MOS. A category "A" MOS identifies the primary skill/knowledge of a Marine. "A" billets are staffed with Marines whose primary/additional MOS is the same as the MOS that identifies the billet; e.g., the billet of Tank Crewman is staffed by a Marine whose primary/additional MOS is 1812.

b. "B" billets are identified by a category "B" MOS. These MOS's designate a particular skill/training possessed by Marines that is in addition to their PMOS. They may be assigned only as skill designator MOS's; e.g., Recruiter 8411, Career Planner 8421, and Drill Instructor 8511, etc.

c. Billet designators are identified by a variety of MOS's requiring general skills. These billets may be staffed by Marines with any PMOS and require solely for that purpose and will not be assigned as an MOS to any Marine. Examples of billet designators and MOS's are Guard 8151, Food Service Attendant 8915, and Unrestricted Officer 9910.

2. Every Marine can reasonably expect to be considered for assignment to a "B" billet at some time within their career. The allocation of available manpower spaces to MOS's and grades includes "A" billet requirements, "B" billet requirements, and overhead transients, trainees, prisoners, and patients (T2P2)).

Military

MCO P1300.8, *Marine Corps Personnel Assignment Policy.* Navy personnel assigned to FMF units will be subject to the tour lengths prescribed in the current edition of NavPers 15559 for officer personnel, and NavPers 15909 for enlisted personnel.

Officer

MCO P1300.8, *Marine Corps Personnel Assignment Policy.* OFFICER ASSIGNMENTS

Monitors make assignments based on the following priorities (listed in order of precedence):

a. Needs of the Marine Corps.

b. MOS/billet variety - command versus staff tour.

c. Availability of the individual.

d. Overseas Control Date (OCD).

e. Seniority.

f. Individual preference.

FIELD GRADE OFFICER COMMAND TOURS

1. Field grade officers assigned at major shore commands will serve a 36 month tour.

2. Field grade officers assigned as commanders of MCAS's will serve a 36-month tour length. Tour lengths less than or greater than these bounds will be approved by the CMC (MM).

Restricted officers (warrant officers and limited duty officers) shall only be assigned to restricted officer billets within their MOS.

1. Field grade officers assigned at major shore commands will serve a 36 month tour.

2. Field grade officers assigned as commanders of MCAS's will serve a 36-month tour length. Tour lengths less than or greater than these bounds will be approved by the CMC (MM).

Key Billet. An overseas position (officer or warrant officer only) of extremely unusual responsibility, where it has been determined that the continued presence of the incumbent is absolutely essential to the mission of the activity or unit or to the United States presence in that area.

MCO P1000.6G, *Assignment, Classification, and Travel System Manual.* 2203. WARRANT OFFICERS

1. Warrant Officers shall be assigned to warrant officer billets within their MOS or, in special situations, to LDO billets within the warrant officer's occupational field where the strength of the LDO MOS is extremely short.

Warrant officers shall not be assigned to billets designated for unrestricted officers.

General Officers. General officers are assigned PMOS 9903 upon acceptance of appointment to the grade of brigadier general. General officers are not normally assigned additional MOS's.

b. Colonels. Unrestricted officers promoted to the grade of colonel are assigned an appropriate PMOS, e.g., 9904, 9906, 9907, or 9914, based upon the PMOS held at the time of promotion. MCO P1200.7, MOS Manual, sets forth which PMOS's/OccFld's are appropriate for each of the aforementioned MOS's designated for assignment to colonels.

An additional MOS will be assigned to an unrestricted officer completing a formal course of instruction that qualifies the officer for a category I/II/III MOS. An additional MOS may be assigned to an unrestricted officer who:

a. Has served in a category I/II/III MOS billet for at least 6 months, and

b. Is certified by competent authority and recommended by the commander/reporting senior.

3. LDO's and warrant officers will be assigned an additional MOS within the same OccFld upon completing a formal course of instruction that qualifies the officer for a category II/III MOS.

2106. SUPPLEMENTARY MOS (SMOS)

1. The SMOS is designed to provide qualified officers to MOS's that have shortages, as defined by CMC (MPP), at the grades of first lieutenant through lieutenant colonel. SMOS's may be assigned voluntarily or involuntarily. The classification may be made by the officer retention Board (ORB), an SMOS selection board, or by the CMC (MMOA) as directed by the Deputy Chief of Staff for Manpower and Reserve Affairs. Service by officers in an SMOS will normally occur following the officer's first FMF tour. Follow on tours are possible when officers obtain field grade rank. The assignment of an SMOS does not affect currently held additional MOS's and will not change an officer's PMOS.

2201. UNRESTRICTED OFFICERS. Unrestricted officers are considered to have a specialization upon which duty assignments are based. Unrestricted officers are rotated between command and staff assignments to enhance their training and experience. Such rotational assignments will be made on the basis of prior service, education, personal preference, and Marine Corps requirements.

2202. LIMITED DUTY OFFICERS. Limited duty officers shall be assigned to limited duty officer billets within their MOS. Limited duty officers shall not be assigned to billets designated for unrestricted officers.

2203. WARRANT OFFICERS

1. Warrant Officers shall be assigned to warrant officer billets within their MOS or, in special situations, to LDO billets within the warrant officer's occupational field where the strength of the LDO MOS is extremely short. Warrant officers shall not be assigned to billets designated for unrestricted officers.

2207. DIRECTED ASSIGNMENTS

1. The CMC may direct the assignment of unrestricted officers to 3-year tours of duty outside their PMOS. The assignments to other OccFld's will be made in response to valid Marine Corps requirements. Officers in this category will be assigned an appropriate additional MOS to identify the OccFld in which they are serving.

MCO 1200.17, *Military Occupational Specialties (MOS) Marine Corps Manual.* Officer Designation Officers are generally assigned primary MOSs by their type of commission or appointment.

a. The three groups of MOSs that are assigned as primary MOSs to officers are: Group I MOSs suitable for assignment to unrestricted Regular/Reserve officers; Group II MOSs suitable for assignment to LDOs; and Group III MOSs suitable for assignment to regular/reserve warrant officers.

Enlisted

MCO P1000.6G, *Assignment, Classification, and Travel System Manual.* 2204. TEMPORARY OFFICERS. During periods of increased operational commitments, selected enlisted personnel may be commissioned as temporary officers to meet expanded officer requirements and individual qualifications.

The distribution system is based on the concept that any Marine qualified in a particular grade and MOS can efficiently perform the duties associated with any billet designated for that same grade and MOS. The description of, and qualifications for, each MOS at the various grade levels are prescribed in the MOS Manual. These descriptions and qualifications pertain both to billets and individual Marines, except where specifically prescribed. When the billet duties actually performed by the Marine are not commensurate with the grade and MOS for the billet designated, the CMC (MP) should be notified. In such cases the billet will be analyzed and redesignated, or it will be directed that the duties performed by the Marine be modified to correspond to the designated grade and MOS.

Enlisted personnel assigned to billets on a military emergency or short-term operational requirement basis (as differentiated from restrictive or retraining assignments) are considered as filling billet vacancies.

The responsibility and authority for the distribution of enlisted Marines among commands essentially parallels that prescribed for commissioned officers. In the case of enlisted Marines, however, assignment to permanent duty stations will always be based on requirements by grade and specialty.

Marine Corps Reserve. Provided a billet vacancy exists and subject to the requirements in the MOS Manual, commanding officers may assign Marines to OJT in other than their PMOS.

3303. RESPONSIBILITY FOR ASSIGNMENT

1. The responsibility and authority for the distribution of enlisted Marines among commands essentially parallels that prescribed for commissioned officers. In the case of enlisted Marines, however, assignment to permanent duty stations will always be based on requirements by grade and specialty.

2. Commanders are responsible for the proper use of Marines assigned to their commands. Unless the Marine is specifically directed to fill a "B" billet or is otherwise directed by higher authority, commanding officers will make every effort to assign Marines in MOS's that are short to billets in their PMOS. To assist commanders in identifying MOS overages/shortages, HQMC periodically publishes Marine Corps bulletins in the 1220 series. This information can be used as the basis for requests for retraining or OJT to correct existing imbalances.

3306. RESTRICTIVE ASSIGNMENTS

1. Once classified in other than a basic MOS, Marines will, wherever possible, be assigned only to billets with the same, equal, or higher MOS in the same OccFld as their PMOS. When this is not possible, priority in assignments will be as follows:

a. To billets with a related MOS; i.e., an assigned additional MOS which falls within the same OccFld as the PMOS.

b. To billets which provide unit training or formal school training for an unassigned MOS appropriate to the grade held, and within the same OccFld as the PMOS.

c. To billets in an assigned additional MOS not necessarily within the same OccFld as the PMOS.

Civilian

Not applicable.

Bibliography

Cavallaro, Gina, "NCO Training Overhaul on the Way," *Army Times*, November 3, 2008.

Chisholm, Patrick, "Learning to Fly . . . UAVs," *Military Training Technology (Online Edition)*, December 20, 2007.

CNO Washington DC (UC), NAVADMIN 067/09, *Active Duty Flying Chief Warrant Officer Pilot Program*, March 2009.

Dahlman, Carl J., *The Cost of a Military Person-Year: A Method for Computing Savings from Force Reductions*, Santa Monica, Calif.: RAND Corporation, MG-598-OSD, 2007. As of August 16, 2010: http://www.rand.org/pubs/monographs/MG598/

Department of Defense, DoDD 1100.4, *Guidance for Manpower Management*, February 12, 2005.

———, DoDI 1100.22, *Guidance for Determining Workforce Mix*, September 7, 2006, Incorporating Change 1, April 6, 2007.

Department of Defense, Under Secretary of Defense for Personnel and Readiness, *Report of the 9th Quadrennial Review of Military Compensation*, Volume 2, March 2002.

Department of Labor, Bureau of Labor Statistics, "Occupational Outlook Handbook," 2008–2009.

———, "Education and Training Data," last modified December 10, 2009. As of August 17, 2010: http://www.bls.gov/emp/ep_data_education_training.htm

Department of the Air Force, AFPD 38-2, *Manpower and Organization*, June 1, 1996.

———, AFI 38-204, *Programming USAF Manpower*, August 1, 1999.

———, AFI 38-201, *Determining Manpower Requirements*, December 30, 2003.

———, AFM 36-2108, *Enlisted Classification*, October 31, 2004.

———, AFI 36-2101, *Classifying Military Personnel*, June 14, 2010.

Department of the Army, AR 71-32, *Force Development and Documentation—Consolidated Policies*, March 3, 1997.

———, AR 611-1, *Military Occupational Classification Structure Development and Implementation*, September 30, 1997.

———, Army Pamphlet 600-25, *U.S. Army Noncommissioned Officer Professional Development Guide*, October 15, 2002.

———, AR 570-4, *Manpower Management*, February 8, 2006.

———, FM 6-22, *Army Leadership—Competent, Confident, Agile*, October 2006.

———, Army Pamphlet 600-3, *Commissioned Officer Professional Development and Career Management*, 2007.

———, Army Pamphlet 611-21, *Military Occupational Classification and Structure*, January 22, 2007.

Department of the Navy, *U.S. Navy Regulations 1990, Precedence, Authority and Command 1001, Officers of the Naval Service*, September 14, 1990.

———, OPNAVINST 1210.2B, *Navy Officer Occupational Classification System (NOOCS)*, August 16, 1993.

———, OPNAVINST 1223.1B, *Navy Enlisted Occupational Classification System (NEOCS)*, August 16, 1993.

———, MCO P1300.8R, *Marine Corps Personnel Assignment Policy*, October 4, 1994.

———, MCO P1000.6G, *Assignment, Classification, and Travel System Manual*, May 6, 1999.

———, OPNAVINST 1001.24, *Individual Augmentation (IA) Policy and Procedures*, July 5, 2000.

———, OPNAVINST 1420.1A, *Enlisted to Officer Commissioning Programs Application Administrative Manual*, June 7, 2004.

———, SECNAVINST 1000.7F, *Interservice Transfer of Commissioned Officers*, January 5, 2006.

———, SECNAVINST 1412.9B, *Marine Corps Limited Duty Officer and Warrant Officer Programs, Promotions, and Continuation Procedures*, February 7, 2006.

———, OPNAVINST 1000.16K, *Navy Total Force Manpower Policies Procedures*, August 22, 2007.

———, MCO 1200.17, *Military Occupational Specialties (MOS) Marine Corps Manual*, May 23, 2008.

———, NAVADMIN 067/09, *Active Duty Flying Chief Warrant Officer Pilot Program*,

———, NAVPERS 18068F, *Manual of Navy Enlisted Manpower and Personnel Classifications and Occupational Standards*, Vol. II: *Navy Enlisted Classifications (NECs)*, July 2010.

Eitelberg, Mark J., *Manpower for Military Occupations*, Alexandria, Va.: Human Resources Research Organization, 1988.

Future Force, "HRStory: The Past, Present, and Future of Human Resources in the U.S. Coast Guard," November 2003.

Gottschalk, Peter, and Michael Hansen, "Is the Proportion of College Workers in Noncollege Jobs Increasing?" *Journal of Labor Economics*, Vol. 21, No. 2, April 2003, pp. 449–471.

Hansen, Michael, and Jennie Wenger, *Is Enlisted Retention Too High?* Alexandria, Va.: CNA, CNA Research Memorandum D0008594, October 2003.

Horowitz, Stanley A., *Skill Mix, Experience, and Readiness*, Alexandria, Va.: Center for Naval Analyses, October 1983.

Joint Chiefs of Staff, CJCSI 1805.01, *Enlisted Professional Military Education Policy*, October 28, 2005.

———, CJCSI 1001.01, *Joint Manpower and Personnel Program*, December 28, 2004, Incorporating Change 1, March 13, 2008.

———, CJCSI 1330.01D, *Assignment of Officers (O-6 and Below) to the Joint Staff*, April 1, 2008.

———, CJCSI 1330.05, *Joint Officer Management Program Procedures*, May 1, 2008.

———, CJCSI 1320.01D, *Assignment of Enlisted Personnel to the Joint Staff*, May 23, 2008.

Kirby, Sheila Nataraj, and Harry J. Thie, *Enlisted Personnel Management: A Historical Perspective*, Santa Monica, Calif.: RAND Corporation, MR-755-OSD, 1996. As of August 16, 2010: http://www.rand.org/pubs/monograph_reports/MR755/

Murnan, Agneta, "Student Pilot Gets First UAS Assignment at Vance," *Air Force Link*, October 23, 2008.

Navy Manpower Analysis Center, *Ship/Fleet Manpower Document Development Procedures Manual*, January 2005.

Office of Personnel Management, "Federal Classification and Job Grading Systems," undated. As of August 17, 2010:
http://www.opm.gov/fedclass/index.asp

Public Law 96-513, 94 Stat. 2835, The Defense Officer Personnel Management Act, December 1980.

Rostker, Bernard, *I Want You: The Evolution of the All Volunteer Force: History and Analysis*, Santa Monica, Calif.: RAND Corporation, MG-265-RC, 2006. As of August 16, 2010:
http://www.rand.org/pubs/monographs/MG265/

Thie, Harry J., Roger Allen Brown, Mark Berends, Rudolph H. Ehrenberg, Ann Flanagan, Claire M. Levy, William W. Taylor, Richard Eisenman, William Fedorochko, Clifford M. Graf II, Mark Hoyer, Paul Bracken, Norman T. O'Meara, Jerry M. Sollinger, Judith Larson, Kathy Mills, Nancy L. Rizor, Bernard D. Rostker, Glenn A. Gotz, Charles Robert Roll, Jr., and Colin O. Halvorson, *Future Career Management Systems for U.S. Military Officers*, Santa Monica, Calif.: RAND Corporation, MR-470-OSD, 1994. As of August 16, 2010:
http://www.rand.org/pubs/monograph_reports/MR470/

Thie, Harry J., John Christian, Matthew Stafford, Roland J. Yardley, and Peter Schirmer, *Fiscally Informed Total Force Manpower*, Santa Monica, Calif.: RAND Corporation, MG-606-OSD, 2008. As of August 16, 2010:
http://www.rand.org/pubs/monographs/MG606/

United States Code, Title 10, Armed Forces, February 1, 2010.